AF839659

Dychweler y llyfr hwn erbyn y dyddiad isod, neu pan elwir amdano gan y Llyfrgellydd.
Dirwyir unrhyw ddarllenydd a fetho ddychwelyd y llyfr mewn pryd.

This book must be returned on or before the date shown below, or when required by the Librarian.
A fine will be imposed on any borrower failing to return the book in time.

TIME AND CAUSE

PHILOSOPHICAL STUDIES SERIES IN PHILOSOPHY

Editors:

WILFRID SELLARS, *University of Pittsburgh*
KEITH LEHRER, *University of Arizona*

Board of Consulting Editors:

JONATHAN BENNETT, *Syracuse University*
ALAN GIBBARD, *University of Pittsburgh*
ROBERT STALNAKER, *Cornell University*
ROBERT G. TURNBULL, *Ohio State University*

VOLUME 19

RICHARD TAYLOR

TIME AND CAUSE

Essays Presented to Richard Taylor

Edited by

PETER VAN INWAGEN

Syracuse University, Department of Philosophy, College of Arts and Sciences, Syracuse, N.Y., U.S.A.

D. REIDEL PUBLISHING COMPANY

DORDRECHT : HOLLAND / BOSTON : U.S.A.

LONDON : ENGLAND

Library of Congress Cataloging in Publication Data

CIP

Main entry under title:

Time and cause.

(Philosophical studies series in philosophy ; v. 19)
'Bibliography of Richard Taylor': p.
Includes bibliographical references and index.
1. Causation – Addresses, essays, lectures. 2. Time – Addresses, essays, lectures. 3. Philosophy – Addresses, essays, lectures. 4. Taylor, Richard, 1919– – Addresses, essays, lectures. I. Taylor, Richard, 1919– II. Van Inwagen, Peter.
BD591.T55 122 79-27401
ISBN 90-277-1048-1

Published by D. Reidel Publishing Company,
P.O. Box 17, 3300 AA Dordrecht, Holland

Sold and distributed in the U.S.A. and Canada
by Kluwer Boston Inc., Lincoln Building,
160 Old Derby Street, Hingham, MA 02043, U.S.A.

In all other countries, sold and distributed
by Kluwer Academic Publishers Group,
P.O. Box 322, 3300 AH Dordrecht, Holland

D. Reidel Publishing Company is a member of the Kluwer Group

BD591.T5
348070
STACK

All Rights Reserved
Copyright © 1980 by D. Reidel Publishing Company, Dordrecht, Holland
No part of the material protected by this copyright notice may be reproduced or utilized in any form or by any means, electronic or mechanical, including photocopying, recording or by any informational storage and retrieval system, without written permission from the copyright owner

Printed in The Netherlands

TABLE OF CONTENTS

PREFACE

Richard Taylor was born in Charlotte, Michigan on 5 November 1919. He received his A.B. from the University of Illinois in 1941, his M.A. from Oberlin College in 1947, and his Ph.D. from Brown University in 1951. He has been William H. P. Faunce Professor of Philosophy at Brown University, Professor of Philosophy (Graduate Faculties) at Columbia University, and Professor of Philosophy at the University of Rochester. He is the author of about fifty articles and of five philosophical books.

This volume consists of essays presented to Richard Taylor on the occasion of his sixtieth birthday. Some of the contributors have been Taylor's students; some have been his colleagues; and all have been, and continue to be, his admirers. I have made several attempts to articulate what it is I (I would not presume to speak for anyone else) admire about Richard Taylor:

(1) There is a particular 'flavor' to Taylor's philosophical writing and conversation that is wholly delightful. Like any other flavor, it can be tasted and enjoyed and remembered but never adequately described. (If there should be someone who has picked up this book who does not know what I mean, I recommend that he read the chapter on 'God' in Taylor's *Metaphysics*.)

(2) Taylor is a masterful dialectician. Anyone who has spent much time talking about philosophical questions with him is apt to have discovered that his own arguments, if sound, would prove far more than he could bring himself to believe. Most of us, I suppose, would cheerfully assent to Schopenhauer's dictum that an argument is unlike a cab in that one cannot dismiss an argument when it has taken one as far as one wants to go. A philosophical conversation with Richard Taylor, however, has a disconcerting way of making one see that this dictum applies to oneself.

(3) Insofar as this lies within any man's power, Taylor is free from the illusions of the present age. (I use 'illusion' in Freud's sense: an illusory belief may be *true* for all it is illusory.) Perhaps the following quotation will show that I mean:

> It is . . . quite uncritically thought to be 'scientific' to hold that men are ('ultimately') very like machines and that their behavior can ('ultimately', as we are always told) be understood in terms of the same principles by which the behavior of inanimate things is understood. Or, again, it is somehow deemed 'scientific' to deny that men ever act freely.

Hence arguments in favor of the causal determinism of human behavior are always received with the keenest interest and guaranteed an audience, while arguments casting doubt upon this hypothesis are generally met with scepticism and even hostility. Arguments of the former kind need not even be very good, philosophically. They can be question-begging, or even quite irrelevant, like so many of the speculations of psychologists. They are nevertheless sensed to be somehow 'on the right track' – scientific in spirit if not in content. Of course, their being on the right track does not in the least entail their being cogent, objective, undogmatic, or philosophically perceptive. Instead, it results from their seeming somehow to fit more or less into the general conception of what reality must be like, a conception which was borrowed from physical science, and which thus inherited the honorific appelative 'scientific'. (*Action and Purpose*, pp. 5–6.)

But reflections of this sort are not really very effective in conveying what I want to convey. For one thing, they are fragmentary. They give no hint of the way in which the attributes they describe join together to produce a unique philosopher. For another, they are not all the fragments: there are other pieces to the puzzle, pieces I do not know how to begin to describe. With this confession of near failure, I give it up and beg the reader to turn to the essays, which we the contributors to this volume present to Richard Taylor with affection and gratitude.

THE EDITOR

PART I

TIME

J. J. C. SMART

TIME AND BECOMING

Richard Taylor has made valuable contributions to the discussion of many central questions of philosophy. Of particular interest to me has been his work on the philosophy of time. I agree with some of his conclusions and disagree with others, but whether his conclusions are true or false his arguments are always challenging. I find Taylor's arguments for similarities between time and space particularly good, as in his fine papers 'Spatial and Temporal Analogies and the Concept of Identity'[1] and 'Moving About in Time'.[2] On the other hand I do not agree with certain views which Taylor has held about an alleged difference between time and space, namely that there is a passage or movement of time, an irreducible *pure becoming*. I refer here in particular to Chapter 8 of the second edition of Taylor's *Metaphysics.*[3] In this paper I intend to argue for a contrary view, namely that the alleged passage of time or pure becoming is an illusion. Certainly we *feel* that time flows, but I want to say (contrary to Taylor) that this feeling arises out of metaphysical confusion. It is easier, in my opinion, to argue for the illusoriness of temporal passage or of pure becoming than it is to give a plausible diagnosis of the cause of the illusion, but I shall try to make some suggestions about this toward the end of this paper. In defending the ideas of passage and pure becoming Taylor has ranged himself alongside many other philosophers, and I shall be concerned with this tradition in general, and not only with Taylor's own particular version of it. It should be noted that Taylor is himself quite aware of the sorts of arguments which I shall put forward, but he is more impressed than I am by the apparent 'datum', as he calls it, of our experience of time as apparently involving a pure becoming or passage.

It is undeniable that we do commonly talk of the passage or flow of time. We thus think of time as like a river bearing us inexorably into the future toward the big waterfall which is our death. Alternatively we can think of time as like a river flowing past us, while we sit on its bank, so that events are born down from the future toward us, and then recede away from us downstream into the past. When we think in either of these ways we think of the passage or flow of time as an objective feature of the universe, which would occur whether or not we were conscious of it. As Newton said, "Absolute, true and mathematical time, of itself, and from its own nature, flows equably

Peter van Inwagen (ed.), Time and Cause, 3–15.
Copyright © 1980 *by D. Reidel Publishing Company.*

without relation to anything external."[4] Other writers treat passage as the movement of our consciousness through time, so that on this view events are just there in time (or rather space-time) and our consciousness just comes across them in its advance into the future.[5] According to this view the movement or passage belongs to consciousness, not to anything in the physical world. I find the notion of the movement of our consciousness just as incoherent as the movement of time itself (or of events in time). Even if I did not I would find a further difficulty in this treatment of passage as the passage of our consciousness through a nonchanging physical world, because I hold that there is no separation between the physical and the mental anyway. I hold that mental processes are just physical processes in the brain, and so if passage were passage of something mental it would still be a physical phenomenon.

If the passage of time is supposed to be literally a motion, whether of time, or of events in time, or of our consciousness through time (or space-time), then there is *prima facie* a simple objection to the idea. Motion is rate of change of spatial position with respect to temporal position. What then could be meant by the motion of time itself or of motion through time? Would this motion be at the rate of one second per second? Admittedly A. N. Prior has seemed happy with this answer.[6] However what is wrong with it can perhaps be brought out by reminding ourselves about what motion through space is. If you and I are moving with respect to one another, then your world line and mine are inclined at an angle to one another in space-time. If you and I are at rest with respect to one another then our world lines are parallel. But how on earth would one represent a movement of one second per second? It is true, as Prior points out, that after one second I have got older by a second. But equally one could say that a ruler gets larger in a left to right direction (say) by one centimeter per centimeter. There is no notion of 'flow' or 'passage' here.

If one believes, then, in the flow or passage of time, one has to take 'flow' or 'passage' in some metaphorical sense. Indeed what one has to do is to concentrate on what seems to be left out in the notion of 'getting older by a second per second' if this is taken in the sense in which a ruler gets larger by a centimeter per centimeter. What is allegedly left out is 'becoming'. But even this word 'becoming' has to be taken in a nonstandard sense. C. D. Broad has used the term "absolute becoming" and Richard Taylor has the term "pure becoming"[7]. The question is whether one can make any clearer sense of this than one can of passage or flow as simple motion.

Understanding the word 'becoming' as it is ordinarily used does not take us very far toward understanding the notion of 'pure' or 'absolute' becoming. Broad, in his *Scientific Thought,*[8] talked of "events becoming", and 'become' here is evidently used as a monadic predicate. Normally we talk of something becoming something or other, as when we say that a man has become fat, or his hair has become grey. In the pure becoming of an event, what does the event become? Perhaps we can restore syntactic parity with ordinary becoming by saying that the event becomes present. Still, this does not seem very helpful, since every event becomes present at some time or other. So the notion of becoming present seems a pretty empty notion, and this is even more obvious when we recognize the indexical nature of words like 'present', 'past', and 'future'.[9] When a person P utters at a time t the sentence 'Event E is present' his assertion is true if and only if E is at t. More trivially, when P says at t 'time t is now' his assertion is true if and only if t is at t, so that if P says at t 't is now' his assertion is thereby true.

Richard Taylor tries to bring out the distinction between pure becoming and ordinary becoming by contrasting two senses of 'becoming older'.[10] Sometimes when we say that someone has become older, or has aged, we mean to convey that he has become more decrepit, that his skin has become wrinkled, that his hair has become grey, and so on. That is, we imply that a certain sort of change has occurred in the person. Taylor contrasts this with the sense in which something gets older even though it suffers no change whatever. Suppose that a crystal remains quite unchanged for ten years, or at any rate suffers no deterioration. It is still true, however, that the crystal has become ten years older: the crystal after ten years is ten years older than it was ten years before. But where has the notion of passage got to? Consider a north-going highway at milestone 100. The highway here is ten miles more northerly than it is at milestone 90. There is no suggestion of 'passage' or 'pure becoming' in this case. It is not clear that getting older implies passage or pure becoming either. The notion of pure becoming itself remains as elusive to me as ever.

If a person gets older in the sense that his hair has become grey and so on, this means simply that a later temporal stage of him is different in various particular respects, such as hair color, from an earlier temporal stage of him. In the other sense of 'getting older' whereby the mere fact of the so-called passage of time makes a person 'get older', there is also a difference of a sort, apart from the mere difference of temporal position and relations to changing things elsewhere in the universe. This difference, so far as I can see, is simply the difference of identity: the later temporal stage is not identical with the earlier temporal stage. However the relation of identity is a queer or degenerate sort of relation,and the predicate 'is identical with' has good claims to being regarded as belonging to pure first-order logic. Indeed in a first-order language with finitely many predicates the identity predicate is eliminable.[11]

The notion of pure becoming is connected with that of events receding into the past and of events in the future coming back from the future to meet us. This notion seems to me to be unintelligible. What is the 'us' or 'me'? It is not the whole person from birth to death, the total space-time entity. Nor is it any particular temporal stage of the person. A temporal stage for which an event E is future is a different temporal stage from one for which event E is present or past. Nor is the change of E from future through present to past a genuine change. To think that it is a genuine change is to treat the indexical expressions 'past', 'present', and 'future' as though they were non-indexical predicates like 'is red' or 'runs'. Nor do I think that introspection of our temporal experience (which is of course all our experience) can give meaning to such expressions as 'pure becoming' or "the transitory aspect of temporal facts".[12] A concept has to be a public concept or no concept at all: if there is no intersubjective way of singling out the concept of pure becoming, how on earth can I know the aspect of inner experience to which I am supposed to apply the concept?

The elusiveness of the notion of temporal becoming can be brought out if we consider a recent attempt by Storrs McCall to explicate the notion of temporal passage, in his article "Objective Time Flow".[13] In this paper McCall argues that objective time flow is connected with irreducible contingency in the world, so that in a deterministic world there would be no time flow. He compares his view of the universe with the theory of branching universes which is apparently implied by Hugh Everett III's interpretation of quantum mechanics.[14] Consider the situation of Schrödinger's cat paradox. There is an animal trapped in a box together with a device which will trigger off a gun which shoots the cat if and only if some radioactive material decays within a certain time, the probability of the decay being one half. The wave function for the system is a superposition of wave functions giving 'cat dead' and 'cat alive'; on the usual interpretation a 'measurement' (which in this case would be the shooting or nonshooting of the cat) leads to a collapse of the wave function into one or other of the two states. On the Everett interpretation the wave function does not suffer the discontinuous collapse but proceeds deterministically; however the *universe* splits into two branches, one containing a live cat and one a dead cat. Indeed on this view the universe is suffering innumerable such splittings all the time. We can compare the universe to a shrub with many branches, each branch itself having many branches, and so on. McCall's picture of the universe is a modification and elaboration of this sort of thing. He holds that at time *t* the space-time universe is in a state which contains all the branches after *t* but only the trunk before *t*. As time proceeds there is a continual lopping-off of branches on the lower part of the shrub, so as to leave a single trunk. Any branch can become part of a later trunk, and so we can envisage grotesquely bent trunks.[15] It is this continual change of the whole space-time universe which according to McCall constitutes the objective flow of time.

This seems to imply a proliferation not only of branches of the shrub but also a proliferation of shrubs. After all a given shrub either has or has not branches on a certain lower part (from ground level upward) of its trunk. A single space-time universe surely either has branches before *t* or it does not have branches before *t*. We must suppose therefore a vast multiplicity of universes, one for each value of *t*. Think of a universe with branches after *t* but none before *t* as a card with a shrub drawn on it. Then McCall's picture suggests to me that there is a super-universe which is like a pack of continuum-many cards, one above the other, cards higher in the pack portraying a longer unbranched 'trunk' than those lower in the pack. Indeed McCall has to complicate his theory in order to deal with special relativity, because of the relativity of simultaneity: the universe branches 'now', let us suppose, but there is no unique 'now'. So we must (to continue our analogy) have continuum-many piles of continuum-many cards, each pile corresponding to a different time direction (or set of inertial axes). There is a super-universe consisting of continuum-many arrays of continuum-many branched universes. I think that McCall does not want to commit himself to this huge ontology. He says that "the analogy with a three-dimensional tree, which grows and changes in time, should not be carried too far", and goes on to say that we can "deny that every entity which is different at different times changes", or alternatively that we can say that "the progressive falling away of future branches on the universe-tree does not 'take' time, but instead 'generates' time".[16] It is unclear to me how the first alternative helps, and the second one seems to me to be quite obscure, and so I can not see how McCall can avoid the huge ontology. However it may be the case that a 'Heraclitean' will find things quite intelligible which are quite obscure to a 'Parmenidean' like me.

In correspondence McCall has kindly commented on my interpretation of him, and has said that in one important respect it misses the spirit of what he intended to convey. He holds that the universe at time *t* is not a slice of some super-universe, something analogous to a card in a deck. The universe at time *t* consists of just the universe at *t*, and the universe at t', where t' is earlier or later than at t, does not exist at all. My worry is this: if the universe *now* is an entity, how can the universe at some other time be a nonentity? After all, McCall seems to be able to say things about it.

McCall further remarks that the deck of cards could not include a card for (say) a time in 1999, since nothing now indicates which branches of the universe should be lopped off. From my point of view this is too verificationist. We may not be able to tell what branches would be going to be lopped off, but the ones which would be lopped off are the ones which would be lopped off! What will be will be! No doubt once more, I am too Parmenidean to understand the Heraclitean point of view. I can approximate to doing so only by a theory of a super-universe which the Heraclitean McCall naturally rejects. I must leave it to the reader to judge whether this rejection is intelligible.

In our ordinary theory of space-time, change is a matter of one time slice or temporal stage of the space-time universe being different from an earlier time slice. McCall's theory is concerned to explicate the queer sort of change which is implied by time flow. As far as I can see this would be a matter of one 'super-universe slice' being different from another 'super-universe slice', or, on our analogy, one card on a pile being different from another card on the pile.

The lopping-off of branches with time is supposed to do justice to something which McCall and others have thought characteristic of the flow of time. The flow of time is supposed to reduce alternative possibilities to one. At a given time there is only one past but there are alternative futures. Thus there are no branches coming from the past. Branches in the future are supposed to imply a real contingency in the universe. If the universe were fully deterministic then there would be only a single future, just as there is only a single past.

Thus according to McCall's theory, if the universe were strictly deterministic there could be no objective time flow. Now if McCall's theory is mainly to explain our tendency to think that time flows, this is very odd. The human brain probably works fairly nearly as a deterministic machine, because neurons (and even protein molecules and the like) are macroscopic objects as far as quantum mechanics is concerned. It is hard to see how the question of physical determinism or otherwise can affect our subjective feelings about time. Thus it seems to me that we would have the same sort of illusion (as I think it is) of the flow of time even in a deterministic universe. I do not therefore see how McCall's theory of time flow as real and objective (but dependent on contingency) could provide a plausible explanation of this puzzling feature of our temporal experience.

If we confine ourselves to asserting only that an ordinary nonbranching space-time exists, we can of course still accept contingency. To say that the universe is not deterministic is to say that the laws of nature do not connect an event E with some time-slice of the universe earlier than E. The event E of course exists determinately in the future, even though no conceivable knowledge about the present would enable even an infinitely powerful calculator to predict E.

McCall's theory makes the future *bigger* than we commonly believe it is. (It has all those branches.) And of course the whole super-universe, as I have described it in order to make sense of it, consists of continuum-many sets of continuum-many sets of such branched universes. Considerations of parsimony surely make the theory very implausible. It may be instructive to compare McCall's theory with an earlier theory due to C. D. Broad in his *Scientific Thought.*[17] In this theory (one of several theories of time which Broad held during his life) the future is *less* big than I think it is. Indeed according to this theory of Broad's, the future is empty and the world increases in size as time passes and more and more layers of the present get added to it. The past keeps pushing up into the emptiness of the future. At a time t_2 there is *more* universe than there is at an earlier time t_1. The incoherence in Broad's theory comes out when we reflect that this accretion through time of the contents of the universe cannot be a process in time, because this would once more bring us back to the idea of a rate of change of time with respect to time itself. Would the universe get bigger by one second per second, in any sense other than that in which a ruler gets longer by one centimeter per centimeter? Later in his life Broad himself came to reject the notion of temporal flow as motion[18] but he gives no clear idea of what else he thought it might be. One objection which he then made to his theory of accretion of existence was that it suggested that past stages of the universe 'coexist' with present ones. This seems to be a mistaken objection. If 'coexist' means 'coexist *now*' then we can reply that past and present stages do not coexist *now* but they coexist (using this verb tenselessly) and of course they coexist at different times. However according to Broads's theory there would have to be *a* sense in which events at times earlier than now would have to coexist now. Just as I interpreted McCall's theory as implying that there would have to be a super-universe consisting of continuum-many branched universes, 'later' branched universes having longer unbranched parts of their 'trunks' than earlier ones, so I must interpret Broad's picture as a whole continuum-many array of universes so that 'later' ones have more content than 'earlier' ones. Indeed if we modified Broad's theory so as to take account of the relativity of simultaneity, it would consist of continuum-many stacks of continuum-many such worlds, just as in attempting to interpret McCall we thought of the super-universe as analogous to continuum-many stacks of continuum-many cards. So though if we take only *one* of the worlds I have attributed to McCall and one of Broad's then McCall's theory seems to make the universe too big and Broad's makes the universe too small (by leaving out the future); in fact Broad's theory nevertheless ends up (as I interpret it anyway) by implying that there are many worlds and so foists on us a bloated super-universe. It is rather like David K. Lewis's realistic interpretation of possible worlds.[19] In order to make sense of a dubious notion of modality we end up with an unacceptable metaphysics. (This is not to say that all modal notions are unclear: sometimes a metalinguistic account or something of the sort will do the trick.) Similarly, in order to make sense of a dubious notion of pure becoming we end up by postulating a bloated universe. We would do better to reject this dubious notion of modality and the dubious notion of pure becoming.

Neverthless it does *seem* that time passes. When we are bored it seems to pass very slowly. For a criminal awaiting execution it no doubt seems to pass very quickly. Richard Taylor remarks that the criminal's terror at the thought of the rapid diminution of what little is left of his life would hardly be relieved by metaphysical arguments about the absurdity of this sort of change.[20] Perhaps, however, an opponent of the notion of pure becoming could give *some* rational basis to the criminal's terror: he has at t_2 less of his life to make plans about than he has at t_1, where t_2 is later than t_1, and we have a strong biologically explicable propensity to plan for the future. Frustration of this propensity is upsetting to us. Still, no doubt this gives only *part* of the explanation of the criminal's terror.

If the passage of time is an illusion it is a strange and intellectually worrying one. It would be good if we could not only give reasons for thinking that it *is* an illusion (as I have tried to do) but also if we could give some sort of explanation of how this illusion arises. The difficulty is that it seems to be the illusion of an absurdity. A straight stick may look bent, but there is nothing absurd about a bent stick. Perhaps the case is more like that of those impossible pictures which have been drawn by M. C. Escher. In the case of an impossible picture we have the following situation. One part of the picture is a two-dimensional pattern which can be interpreted by the rules of perspective as representing a certain three-dimensional entity. An adjoining part of the picture can be interpreted as representing another three-dimensional entity. Unfortunately one single interpretation of the two-dimensional pattern will not do for both parts: the two parts cannot be fitted together so that the whole two-dimensional pattern represents a single three-dimensional entity. In the middle part of the pattern, where the two parts come together we are forced into trying two inconsistent interpretations at once. Is there anything analogous to this in the genesis of the idea of temporal passage or pure becoming?

Many years ago[21] I held that the illusion of the flow of time came from confusion about indexical expressions such as 'past', 'present', 'future', 'now', and tensed verbs. If we forget the indexical character of the words 'past', 'present' and 'future' we may think that events really change in respect of being future and then present and then past. If a man says in his youth in 1755 'It is ten years since the '45 rebellion' and then in his old age in 1805 he says 'tis sixty years since'[22] he may be tempted to think that there has been a real change of the '45, a recession into the past, whereas all there really is to the matter is the relations of being ten years earlier than one of his utterances and being sixty years earlier than another of his utterances. The events have not changed as a continuant changes: we do not mean that one later temporal stage of the event is different from an earlier temporal stage of it. Indeed we may be tempted to think of even a momentary event as changing from being future to being past, and a momentary event has no distinct earlier and later stages. So the trouble may come from trying to talk of events in terms appropriate only to continuants, and from talking of words like 'future', 'past' and the tenses as though they were nonindexical predicates. Taylor's criticisms of "the attempts to expurgate pure becoming"[23] seem to me to turn on the impossibility of translating indexical expressions, such as tenses, into non-indexical ones. I agree on the impossibility, but I challenge its metaphysical significance, since the semantics of indexical expressions can be expressed in a tenseless metalanguage.

However I have become less sure of this 'linguistic' explanation of the source of the notion of the illusion of passage.[24] The very etymology of the word 'past' suggests that our concepts of past and future are already infected with the notion of passage, rather than the other way round. Of course etymology is a dubious guide in philosophy. Indeed tenses alone are enough to set up the idea of passage: 'was, is, and will be' serves as well as 'was future, is present, and will be past'. Nevertheless it is natural to wonder whether an exotic people who spoke a tenseless language with no indexicals, or perhaps with the single indexical 'this utterance',[25] would not have the feeling that time passes. There are of course obscurities here. There are obvious problems about radical translation of the exotic language. If Quine is right about the indeterminacy of translation there might be alternative translations of an exotic sentence, in the first place as 'time flows', and in the second place as something metaphysically harmless. In which case there might be no 'fact of the matter', as Quine might put it, as to whether or not the exotic people felt that time passes. It is also worth considering whether a person with no language at all could feel that time flows. As Wittgenstein said, 'Can a dog hope?'. Incidentally the case of our exotic people is one in which the linguistic principle of charity need not be applied: in the case of a widespread human illusion it might be better, though uncharitable, to attribute the illusion to the exotic people too. It might even be that the exotic metaphysicians could be most plausibly interpreted (in the process of radical translation) as espousing a philosophy like Bergson's, and this would give us additional reason for translating some of their sentences as 'time flows' and the like.

Let us now explore the possibility of nonlinguistic explanations of the illusion of passage. Remember Broad's theory of the universe getting bigger all the time, as more and more events come into existence. The theory is objectionable, not only because, as I suggested, it involves the notion of movement through time, but also because it makes an unjustified ontological distinction between future and past. Just as a past event *E did* exist, so a future event *F will* exist. Still, it is an empirical fact that people often confuse *X*'s with our experience of *X*'s. Some works on time perception by psychologists have seemed to me to suffer from this defect. So if we say that we feel that time passes perhaps we misdescribe our feeling. Perhaps our feeling is that our *experience* changes. It is not that at t_2 there is more of the *world* than there is at t_1. It is that at t_2 there is more of our *experience* recorded in our memories than there is at t_1. In other words the stock of our memories is continually increasing. At the age of fifty a man has more memories than he has at the age of forty.

But then why do we feel that the world is getting bigger with time rather than getting smaller? In one time direction the stock of our memories is increasing, whereas in the other time direction it is decreasing. It is true that there is a temporal asymmetry about memory. We remember only our previous experiences, whereas we have only theoretical knowledge of our future experiences. Thus I know immediately at 11.00 A.M. that I have memories which I did not have at 10.50 A.M., but I do not know immediately (by memory) that I will have experiences at 11.10 A.M. that I have not had at 11.00 A.M. What I do believe about my experiences at some future time is based on rather general grounds, and is on the assumption that I will not drop dead or become unconscious before 11.10 A.M. But it is hard to see how this temporal asymmetry explains why we feel that the flow of time is in the direction of the increasing stock of our memories.

Another difficulty in the idea that the feeling of the flow of time arises from the increase in our memories comes out if we consider the case of a very old man, who might be losing memories faster than he is gaining them. (We can think here of his long-term memories: presumably information would be going into and coming out of his short-term memory at about the same rate.) I write in ignorance of any psychological literature on the subject, but it seems to me that such a case might well exist, and I would conjecture that nevertheless the old man would still feel that time flows in the same direction in which we feel it does. This throws doubt on the hypothesis that the feeling of the flow of time does after all depend on the increase of our memories.

Adolf Grünbaum has suggested another way in which memory may explain our feeling of the passage of time. He has remarked that "The flux of time consists in the *instantaneous awareness* of *both* the temporal order *and* the *diversity* of the set of the membership of the remembered (recorded) or forgotten events, awarenesses in each of which the instant of its own occurrence constitutes a *distinguished element*".[26] Grünbaum also makes Kant's point that such an awareness is an awareness of succession, not just a succession of awarenesses. But the problem still arises why this should give a feeling not just of a directed temporal order but also one of a temporal *flow* or *passage*. Consider a nested set of arrows, each pointing to the right. If we contemplate it this does not give us a feeling of spatial flow.

Henry Krips has ingeniously pointed out to me, in discussion, another way in which our memories might explain our feeling of temporal passage. If we are floating down a river on a raft we have memories of the places which are upstream from us but not of things which are downstream from us. So 'past' feels a bit like 'upstream', and 'future' feels a bit like 'downstream'.

I am myself inclined to favor yet another suggestion. This is that we are aware of the flow of information through our short-term memories and we confuse this with a flow of time itself. This conjecture is perhaps supported by empirical evidence according to which the greater the number of stimuli that there are in a given temporal interval the greater is the subjective estimate of the length of that interval. It also might be further supported by some more equivocal evidence which suggests that estimation of the length of an interval depends in the same way on the complexity of the stimuli.[27]

However it is still obscure to me how such a flux of information should be misperceived as a flux of time itself.

Australian National University

NOTES

[1] *Journal of Philosophy* **52** (1955), 599–612.
[2] *Philosophical Quarterly* 9 (1959), 289–301.
[3] *Metaphysics*, 2 nd ed. (rev.), Prentice-Hall, Englewood Cliffs, N. J., 1974.
[4] F. Cajori (ed.), *Sir Isaac Newton's Mathematical Principles of Natural Philosophy and His System of the World*, University of California Press, Berkeley, 1960. See the 'Scholium to the Definitions'.
[5] For example, see Sir Arthur Eddington, *Space, Time and Gravitation*, Harper: New York, 1959, p. 51.
[6] See A. N. Prior, 'Time After Time', *Mind* 67 (1958), 244–246.
[7] See C. D. Broad, *Scientific Thought*, Routledge and Kegan Paul, London, 1949, Chap. 2; and Richard Taylor, *Metaphysics*, Chap. 8.
[8] *Op. cit.*
[9] For the semantics of indexical expressions, see Donald Davidson, 'Truth and Meaning', *Synthese* 17 (1967), 304–323, especially the remarks on pp. 319–320.
[10] *Metaphysics*, p. 83 ff.
[11] On this point see W. V. Quine, *Philosophy of Logic*, Prentice-Hall, Englewood Cliffs, N. J., 1970, pp. 63–64.
[12] C. D. Broad, *An Examination of McTaggart's Philosophy*, Cambridge University Press, London, 1938, Vol. 2. Pt. 1.
[13] *Philosophy of Science* 43 (1976), 337–362. I apologize to McCall if I have misunderstood him, as is quite likely in view of my own inability to make sense of theories of time flow. At least I am trying to say what his theory comes to, to the extent that I can make sense of it to myself.
[14] See Bryce DeWitt and Neill Graham (eds.), *The Many-Worlds Interpretation of Quantum Mechanics*, Princeton University Press, Princeton, N. J., 1973. For some cautions see Brian Skyrms, 'Possible Worlds, Physics and Metaphysics', *Philosophical Studies* **30** (1976), 323–332.
[15] See McCall, *op. cit.*, p. 343, bottom diagram.
[16] *Ibid.*, p. 348.
[17] *Op. cit.*
[18] See Broad's reply to his critics, in P. A. Schilpp (ed.), *The Philosophy of C. D. Broad*, Tudor, New York, 1959, especially pp. 766–767.
[19] David K. Lewis, *Counterfactuals*, Blackwell, Oxford, 1973, pp. 84–91.
[20] *Metaphysics*, p. 86.
[21] J. J. C. Smart, 'The River of Time', *Mind* **58** (1949), 483–494.
[22] The alternative title of Sir Walter Scott's *Waverley*.
[23] *Metaphysics*, p. 87 ff.
[24] Recently Ferrel Christensen, in his paper 'The Source of the River of Time', *Ratio* **18**

(1976), 131–144, has defended a linguistic explanation of the metaphor of passage. He holds that predicates like 'past' and 'future' are parasitic on adverbial ones and treats tenses as adverbs; that there are in the ordinary sense no temporal relations, so that 'and then' does not signify such a relation. This seems to me to be wrong. I think that ordinary adverbs should be understood in terms of predicates of events, as has been suggested by Donald Davidson. Tenses should be handled differently, by means of a tenseless metalanguage – see Donald Davidson, *op. cit.* 'And then' should be handled by ordinary conjunction and the predicate 'earlier than'. Nevertheless, even if I am right about these points of disagreement with Christensen, the article is of interest for the way in which he claims to trace 'the river of time' to a linguistic source.

[25] So that instead of 'here', 'now', 'you' and 'will be', for example, they would have 'near this utterance', 'simultaneous with this utterance', 'the person to whom this utterance is addressed' and 'is later than this utterance'.

[26] Adolf Grünbaum, 'Carnap's Views on the Foundations of Geometry', in P. A. Schilpp (ed.), *The Philosophy of Rudolf Carnap*, Open Court, La Salle, Ill., 1963, p. 663. See also Adolf Grünbaum, *Philosophical Problems of Space and Time*, 2 nd ed. (enlarged), Reidel, Dordrecht, 1973, p. 325.

[27] On these points, see Robert E. Ornstein, *On the Experience of Time*, Penguin, Harmondsworth, 1969, Chap. 4.

RODERICK M. CHISHOLM

BEGINNINGS AND ENDINGS

I. INTRODUCTION

In this paper, I will illustrate the way in which Franz Brentano applied the techniques of philosophical analysis to one of the most perplexing areas of metaphysics – a set of *aporiae* about motion and rest and about coming into being and passing away. These questions, which were suggested by Aristotle, were discussed at length by philosophers in the scholastic tradition. And they led Mendelssohn – at least, according to the way in which Kant interprets him – to conclude that the soul is immortal. I believe that anyone who is concerned about these questions could profit by considering what Brentano has said about them.

II. THE PLEROSIS OF TEMPORAL BOUNDARIES: PRELIMINARY STATEMENT

I will first set forth Brentano's theory about the 'plerosis' of temporal boundaries. Then I will try to show how this theory may be used to deal with the *aporiae* in question including that which led to Mendelssohn's attempted proof. What Brentano did here was to make certain suggestions and to point the way. In what follows, I will carry out these suggestions and supply details that are left implicit by Brentano. The best statement of Brentano's own views on these topics has recently been published.[1]

The *plerosis* – or *fullness* – of a boundary is a function of the number of directions in which it is a boundary. Thus a boundary within a temporal continuum may be a boundary in one direction (if it is merely the end point of something that is past or if it is merely the beginning point of something that is future). Or it may be a boundary in two directions (if it is both an end point and a beginning point). But the boundary of a spatial continuum is not thus restricted with respect to the number of directions in which it may be a boundary. It may be a boundary in all the directions in which it is capable of being a boundary, or it may be a boundary in only some of these directions. If a boundary – whether temporal or spatial – is a boundary in all the directions in which it is capable of being a boundary, then it exists in "full

Peter van Inwagen (ed.), Time and Cause, 17–25.
Copyright © 1980 *by D. Reidel Publishing Company.*

plerosis"; otherwise it exists only in "partial plerosis". Brentano believes that by means of the concept of plerosis he can speak in a certain sense of the "parts" of a boundary even though the boundary may have no dimensions. It is one thing to speak of the present as being the end of the past and another thing to speak of it as being the beginning of the future.

Let us now consider the first of our problems: If a thing begins to move is there a last moment of its being at rest or a first moment of its being in motion? There cannot be both, for if there were, then there would be a time between the two moments, and at *that* time the thing could be said neither to be at rest nor to be in motion.[2]

The statement of the problem, Brentano would say, correctly recognizes the *impossibility of adjacent points.* But what it fails to recognize is the *possibility of coincident points.*[3] Brentano's solution is to say that at one and the same moment the thing ceases to be at rest and begins to be in motion. The temporal boundary of the thing's being at rest (the end of its being at rest) is the same as the temporal boundary of the thing's being in motion (the beginning of its being in motion), but the boundary is twofold (*zweiteilig*) with respect to its plerosis. The boundary is in half-plerosis (*in halber Plerose*) at rest and in half-plerosis in motion.

One may object: How can the introduction of a new technical term solve the problem? I would suggest that the introduction of a new technical term *can* be used to solve a problem, provided that the technical term can be used to explicate some of the terms in which the problem is originally formulated. I believe this is the case with Brentano's term 'plerosis'.

To see that this is so, let us replace Brentano's plerosis vocabulary by a slightly different terminology

III. REFORMULATION OF THE DOCTRINE OF PLEROSIS

Consider a thing that is in motion and consider any point or instant of time that falls within the period that the thing is in motion. We should not say that there is motion 'in an instant' – if this is taken to imply that a change can take place in an instant. Yet an instant may fall *within* a time during which something is moving.

If an instant thus falls within a time that a thing A is moving, then: either (a) the instant bounds a prior but not a subsequent motion of A, or (b) the instant bounds a subsequent but not a prior motion of A, or (c) the instant bounds both a prior and a subsequent motion of A. These concepts may be explicated as follows:

> *t* bounds a prior motion of $A =_{df}$. There is a time t' prior to *t* which is such that, for any two periods of time between *t* and t', there is a place which is such that *A* is at that place within one of those periods of time and is not at that place within the other.
>
> *t* bounds a subsequent motion of $A =_{df}$. There is a time t' subsequent to *t* which is such that for any two periods of time between *t* and t', there is a place which is such that *A* is at that place within one of those periods of time and is not at that place within the other.

Aristotle notes that

> there cannot be any irreducible period of time which in its entirety is the 'first' period of the charge, since there is no limit to the divisions of a period, and so you can always show that the change was already taking place before the whole of any period, however minute, has passed.[4]

And he concludes, similarly, that "being brought to rest is an experience that occupies time."[5]

But given the above analysis, we may say what it is for a point of time – a temporal boundary – to be the point at which motion begins, and we may say what it is for a point of time to be the point at which motion ends.

> *A* begins to move at $t =_{df}$. *t* bounds a subsequent but not a prior motion of *A*.
>
> *A* ceases to move at $t =_{df}$. *t* bounds a prior but not a subsequent motion of *A*.

Our definitions enable us to say that the time at which a thing *begins to move* is the same as the time at which it *ceases to rest,* and that the time at which a thing *begins to rest* is the same as that at which it *ceases to move.* And these statements are not contradictory.

One may ask: "If the thing both begins to move and also ceases to rest at *t*, then is it in motion at *t* or is it at rest at *t*? And if it begins to rest at *t* and also ceases to move at *t*, then is it in motion at *t* or is it at rest at *t*? We find the answer to these questions if we say what it is for a thing to move at an instant *t*:

> *A* is in motion at an instant $t =_{df}$. Either *t* bounds a subsequent motion of *A*, or *t* bounds a prior motion of *A*.

(No paradox is involved in thus saying that a thing is in motion at an instant, for our statement does not imply that the thing undergoes change in an instant.) The answer to our question now becomes: at the instant at which the thing begins to move and also ceases to rest, the thing is in motion. And at the instant at which it ceases to move and begins to rest, it is also in motion.

IV. APPLICATION TO A FURTHER PROBLEM

The second of our problems may be put as follows: suppose a body is thrown straight up in the air and is prevented from deviating from its path. Is it at rest for a moment between the time it stops going up and starts to go down (And if so how long?) or does it start to go down at the same moment it stops going up?

Brentano writes: "We should not assume that there is here a moment of complete rest. There is, rather, a moment wherein the beginning of the object's fall coincides with the cessation of its going up."[6] He says that at that moment the motion of the thing is "in half-plerosis" with respect to going up and "in half-plerosis" with respect to going down.

If our suggestion above is correct, we could rephrase Brentano's solution this way. When we say, of any moment, that the thing is going up at that moment, then we leave open three possibilities for that moment: either (a) it bounds both a prior and a subsequent upward movement; or (b) it bounds a prior but not a subsequent upward movement; or (c) it bounds a subsequent but not a prior upward movement – and analogously for going down.

Thus we may distinguish six different situations that obtain during the motion of the thing in question. (1) There is a moment which bounds a subsequent motion of the thing but not a prior motion of the thing; this is the moment at which the thing begins to move. (2) There are moments which bound both prior and subsequent upward motions; these will occur after the moment at which the thing begins to move and before it reaches its highest point. (3) There is a moment which bounds a prior but not a subsequent *upward* motion; this will be the moment at which the thing reaches the highest point. (4) There is a moment which bounds a subsequent but not a prior *downward* motion; this will be the moment at which the thing leaves the highest point. (5) There are moments which bound both prior and quent downward motion; these will occur after the moment the thing leaves the highest point and before it stops moving. And (6) there is a moment which bounds a prior but not a subsequent downward motion: this is the moment at which it ceases to move.[7]

Brentano's suggestion may now be construed as telling us that the moment of (3) and the moment of (4) are one and the same. The moment at which the thing changes its course is one that bounds a prior upward motion and a subsequent downward motion.

Are we saying, then, that when the thing is at the apex it is going up and going down at one and the same time? The answer requires that we make a distinction. If, in saying that something is 'going up at a certain moment' we mean that the moment bounds a prior as well as a subsequent motion, then we should *not* say that the thing is going up at that moment; and analogously for 'going down'. But if, in saying that a thing is 'going up at a certain moment', we mean that the moment bounds either a prior motion or a subsequent motion, then we *may* say that the thing is going up at that moment; and analogously for 'going down'. Hence if we take 'going up' and 'going down' in the first sense, then we should say that when the thing is at the apex, it is neither going up nor going down. But if we take them in the second sense then we may, without contradiction, say that the thing is both going up and going down.

V. MENDELSSOHN'S ATTEMPTED PROOF

Brentano's solution applies in an obvious way to our third problem, which is the source of Mendelssohn's attempted proof of the immortality of the soul. Kant said of Mendelssohn:

> His argument is that since the soul cannot be diminished, and so gradually lose something of its existence, being by degrees changed into nothing (for since it has no parts, it has no multiplicity in itself), there would be no time between a moment in which it is and another in which it is not – which is impossible.[8]

Kant thought that the way to refute Mendelssohn was to supose that there are degrees of reality and that a substance

> may be changed into nothing, not indeed by dissolution, but by gradual loss *(remission)* of its powers, and so, if I may be permitted the use of the term, by elanguescence.[9]

Brentano saw that Mendelssohn's argument was inconclusive and that Kant's doctrine, with its suggestion that there might be a mean between being and nonbeing, is absurd.[10] Mendelssohn's argument can be dealt with in a much more satisfactory way by means of Brentano's analysis of the nature of boundaries. Let us summarize the problem as follows: if a thing ceases to be, then either there is a last moment of its existence or there is a first moment of its nonexistence, but there cannot be both. Which, then, is there?"[11]

Brentano is able to say that there is *both* a last moment of existence and a first moment of nonexistence; they are one and the same. This may be seen by modifying the definitions we have given above. Those definitions pertain to motion and rest, but they may be adapted to existence and nonexistence. The definitions that follow presuppose the concept of a thing existing throughout a period of time – a stretch of time that is not an instant.

> *t* bounds a prior period of *A*'s existence $=_{\text{df.}}$ There is a time t', which is prior to *t* and such that *A* exists throughout every period of time between *t* and t'.

> *t* bounds a subsequent period of *A*'s existence $=_{\text{df.}}$ There is a time t' which is subsequent to *t* and such that *A* exists throughout every period of time between *t* and t'.

And now we may speak of the time at which *A* begins to exist and of the time at which *A* ceases to exist:

> *A* begins to exist at $t =_{\text{df.}}$ *t* bounds a subsequent but not a prior period of *A*'s existence.

> *A* ceases to exist at $t =_{\text{df.}}$ *t* bounds a prior but not a subsequent period of *A*'s existence.

Our definitions allow us to say that 'the first moment of a thing's existence' is the same as 'the last moment of the thing's first period of nonexistence', and that 'the last moment of a thing's existence' is the same as 'the first moment of the thing's second period of nonexistence'.

One may ask: If an instant *t* is the first moment of a thing's existence and the last moment of its nonexistence, then should we say that the thing exists at *t* or that it does not exist at *t*? Or if *t* is the last moment of a thing's existence and the first moment of its nonexistence, then should we say that the thing exists at *t* or that it does not exist at *t*?

The question is, actually: What does it mean to say of a thing that it *exists at an instant*? Hence the question is like one we might have asked about motion: I know what it is for a thing to move throughout an extended period of time, but what is it for a thing to *move at an instant*? Just as we had characterized motion at an instant by reference to motion through a period of time, so, too, we can characterize existing at an instant by reference to existing through a period of time. Let us say, then, what it is for a thing to exist at an instant *t*:

> *A* exists at an instant $t =_{df.}$ Either *t* bounds a subsequent period of *A*'s existence, or *t* bounds a prior period of *A*'s existence.

And so if *t* is the first moment of the thing's existence and the last moment of its nonexistence, then the thing exists at *t*. And if *t* is the last moment of the thing's existence and the first moment of its nonexistence, then the thing also exists at *t*.

Let us now consider the following objection to the thesis that there cannot be both a last moment of one's existence *and also* a first moment of one's (second period of) nonexistence:

> If, then, we divide all instants of time, whether past, present, or future, into two series – those instants at which Socrates was alive, and those instants at which he was not alive – and leave out of consideration, for the sake of greater simplicity, all those instants before he lived, we see at once, by the simple application of Dedekind's Axiom, that, if Socrates entered into eternal life after his death, there must have been either a last moment of his earthly life *or* a first moment of his eternal life, but not both.[12]

The reply is that 'a moment at which Socrates was alive' only *appears* to be incompatible with 'a moment at which Socrates was not alive'. For the moments at which he was alive are of three different kinds: (i) moments such that he existed up to those moments and also from those moments on; (ii) a moment such that he existed from that moment on, but did not exist from any moment up to that moment; and (iii) a moment such that he existed from a time up to that moment but did not exist from that moment on. And the moments at which he was not alive (the moments of his second period of nonexistence) are of two different kinds: (iv) moments such that Socrates failed to exist up to those moments and also failed to exist from those moments on; (v) a moment such that Socrates existed from a certain time up to that moment but did not exist from that moment on. The moments 'at which Socrates was alive' will include (i), (ii), (iii), and (v): the moments 'at which Socrates was not alive' will include (iv) and (v). Hence the two phrases are not contraries. Therefore Brentano can say of Mendelssohn's proof:

> Kant misses the true weakness of this proof. According to Mendelssohn, it is contradictory to suppose that a simple substance ceases to be, since after the last moment of its existence there can be no first moment of its nonexistence. Mendelssohn does not note that one could prove the impossibility of breaking up a composite substance in a similar

way: after that last moment at which its parts are together there can be no first moment at which they are separated. ... Actually, however, there is no contradiction. A temporal point as boundary of a one-dimensional continuum is twofold [*zweiteilig*] in respect of its plerosis. It is logically possible for something to exist up to and at a given period of time and from that same point on *not* to exist. It ceases to be and begins not to be in the same moment. There is no contradiction in supposing a simple substance to be annihilated (though this is ruled out by the general cosmological law according to which God never lets anything go to waste).[13]

In the lectures on metaphysics that Brentano gave at Würzburg between the years 1867 and 1870, he raised such questions as these: Consider that which is in the process of coming into being – should we say that it exists at the time it is coming into being or that it does not exist at the time it is coming into being?; Consider that which ceases to be – should we say that it exists when it ceases to be or that it does not exist when it ceases to be?

In each case, the alternatives that are available seem to be unacceptable. At the time that he gave these lectures, Brentano was inclined to conclude, strictly speaking there *is* no coming into being and passing away.[14] But in saying this, he did not mean that everything exists eternally. He meant, rather, that although a thing may exist at certain times and not at others, there is no event which is that thing's coming into being, and no event which is that thing's passing away. But given his later concept of plerosis, he could say that the instant at which a thing comes into being is an instant at which it exists and that the instant at which it passes away is also one at which it exists.

Brown University

NOTES

1 Franz Brentano, *Philosophische Untersuchungen zu Raum, Zeit und Kontinuum*, ed. by Stephan Körner and Roderick M. Chisholm, Felix Meiner, Hamburg, 1976.

2 Concerning these problems compare: Norman Kretzmann, 'Socrates is Whiter than Plato Begins to Be White', *Noûs* II (1977), 3 –15; 'Incipit/Desinit', in P. K. Machamer and R. G. Turnbull (eds.), *Motion and Time, Space and Matter*, Ohio State University Press, Columbus, 1976, pp. 101–136. See also the references cited at the end of the first of these works.

3 See Brentano, *Psychology from an Empirical Standpoint*, trans. by Antos C. Rancurello, D. B. Terrell, and Linda L. McAlister, Humanities Press, New York, 1973, p. 357.

4 *Physica* 6. 5. 236a.

5 *Physica* 6. 8. 238b.

6 The quotation is from a brief undated manuscript of Brentano's entitled 'Plerose' and classified in Brentano's *Nachlass* as *Megethologie* 15.

7 Or if we leave open the possibility that the thing bounces and immediately goes up

again, we will say: "There is a moment which bounds a prior but not a subsequent downward motion, and this is the moment at which the thing ceases to go down".

8 *Critique of Pure Reason*, B413–414; N. Kemp Smith edition, Macmillan, London, 1929, pp. 372–373.

9 *Op. cit.*, B414; Kemp Smith, p. 373.

10 See Brentano's *Kategorienlehre*, ed. by Alfred Kastil, Felix Meiner. Hamburg, 1968, pp. 90–97.

11 This problem is discussed in detail by F. Suarez, *Disputaciónes Metafísicas* 50. 2. 16, ed. and trans. by S. Rábade Romea, S. Caballero Sánchez, and A. Puigcervar Zanón, Biblioteca hispánica de filosofía, Madrid, Vol. 7, pp. 144–150.

12 Philip E. B. Jourdain, *The Philosophy of Mr. B*rtr*nd R*ss**ll*, George Allen and Unwin, London, 1918, p. 49.

13 *Kategorienlehre*, p. 93.

14 The discussion appears on pp. 50–52 of Brentano's 'Würzburger Metaphysik Kolleg', classified as M96 among the upublished manuscripts of Brentano.

IRVING THALBERG

FATALISM TOWARD PAST AND FUTURE

Richard Taylor has enlivened various fields of analytical philosophy during the past three decades, especially with his ingenious attacks upon commonly held beliefs. I recall being particularly stimulated to reflection by his challenge to one pair of seeming truisms: our certainty that we no longer have any control over what has already happened; and the complementary assumption that some forthcoming events – notably our own deliberate acts – do remain 'up to us'. Taylor has argued separately for backwards causation, and for a fatalistic view of what is going to occur. Most significantly, he seems able to transform every objection we produce against fatalism into an unwanted rationale for supposing that we can shape bygone events – and *vice versa*.

The question whether a current happening might affect earlier events had been perceptively discussed – by Pears,[1] Dummett[2] and others – before Taylor entered the fray (see his essay with Chisholm).[3] But Taylor's contributions have heated and illuminated this controversy. As to fatalism, Gilbert Ryle[4] and R. D. Bradley[5] had offered penetrating refutations of it; yet Taylor's defenses of that doctrine have certainly revived interest in it.[6] My own goal in looking at both these rather esoteric puzzles is to deepen our understanding of the concepts which figure so prominently in them: the key notions of past and future, possibility, causation, power and powerlessness. Taylor's insightful and provocative argumentation will serve as a goad to make us reexamine these concepts, rather than as a target for niggling criticisms. Accordingly, and in order to depersonalize my dissenting analysis, from now on I shall attribute Taylor's reasoning on these subjects to a nameless past-future fatalist with the acronym 'PFF'.

PFF's outlook is quite forthright. He announces:

> A fatalist . . . thinks of the future in the way we all think of the past . . . [namely] as something settled and fixed . . . We are never tempted to try to modify it. It is not in the least up to us what happened last year, yesterday, or even a moment ago. . . .
>
> No one needs to be convinced that fatalism is the only proper way to view the past. That it is also the proper way to view the future is less obvious, due in part, perhaps, to our vastly greater ignorance of what the future holds.[7]

In response, my central question will be: Is this 'hands-off' attitude really just as reasonable – or perhaps equally unwarranted – toward both past and

Peter van Inwagen (ed.), Time and Cause, 27–47.
Copyright © 1980 *by D. Reidel Publishing Company.*

future? If we reject fatalism regarding what will happen, must we swallow backwards causation? In particular, how can we disprove PPF's 'parity' thesis? Here is another terse formulation of it:

> What has happened cannot be undone. The mere fact that it has happened guarantees this. . . .
>
> Whatever the future might hold, there is nothing anyone can do about it now. . . . The mere fact that it is going to happen guarantees this.[8]

If this analogy between "what has happened" and what "is going to happen" is valid, then PFF is correct to see events generally as "unavoidable", and our efforts to mold them as "totally in vain".[9]

But before we consider this possibility, we should mark one verbal trap that awaits us if we speak too glibly of "the future" and what "is going to happen" Our manner of specifying things to come should not by itself imply that such events cannot be avoided. That would beg the fatalistic question under debate. An example might bring this out. Suppose an experienced terrorist has hidden a stick of TNT under the front door of a bank, and set the firing mechanism to detonate at midnight. In these circumstances, we may allude to 'the future blast'. We may predict: 'Material damage from the explosion will be heavy, but no deaths will result from it'. The pitfall here is that such referential and propositional discourse may be ambiguous. Are we referring to, or forecasting, an explosion that will take place regardless of other circumstances, and however anyone acts, between now and 12 P.M.? Or is our prediction conditional? That is, are we instead saying what will occur *unless* something interferes – a misfire, a sudden rainstorm which soaks the mechanism, a last-minute tip by a police informant to the bomb-detection squad? Positively, our 'unless' might amount to saying that a blast will take place if certain conditions hold until midnight. No doubt our 'unless' clauses and lists of continuing conditions will normally remain tacit, as well as uncompletable. Moreover, the future happening we thus conditionally proclaim will very often occur – and for an obvious reason. No interfering snag will turn up, and the conditions required for a detonation will last until midnight. But despite the occurrence of an event that we unconditionally as well as conditionally forecast, we should carefully distinguish these modes of future-tense discourse. Only the first carries a fatalistic suggestion – that what is going to take place is about to occur, come what may. The conditional mode implies the contrary.

By what criteria should we classify a prognosticator's utterance? One would be to ask the prophet whether she or he anticipates the happening whatever else ensues between now and the appointed hour. With regard to fatalistic talk about "the future" and what "is going to happen", I think my 'conditional'-'unconditional' distinction might be particularly germane. Take the reasoning of PFF that I quoted: "there is nothing anyone can do about it [the future] now. . . . The mere fact that it is going to happen guarantees this". The inference to our powerlessness would seem to be valid only if we accept an unconditional reading of the "fact" that such-and-such "is going to happen". Obviously we should reject this, since our dispute with fatalists is precisely whether this or that event will occur, come what may. Incidentally, past-tense speech appears to lack this 'conditional'-'unconditional' distinction. A journalist's report of last night's bombing could hardly mean: "An explosion damaged the bank at midnight yesterday, unless something between now and then has interfered", or "if present conditions hold until midnight yesterday". But this linguistic asymmetry will not disprove backwards causation.

More of that later. According to PFF, our 'hands-off' attitude toward forthcoming events should be modeled on "the way we all think of the past". Nothing we do, and nothing that occurs after some event has taken place can "modify" it. Nor is any subsequent happening or condition required for the earlier event to have occurred. We can dismiss those trick cases in which the description of an earlier happening contains tacit or overt reference to what ensues. To say that a surgeon's endeavors last week *were* effective, or fruitless, is to imply that his patient afterwards recovered, or else sickened. The surgeon made the incisions he made, and so on, regardless of what the sequel turned out to be. The patient's recovery, or decline, was not needed for those incisions to have been made.

PFF would be right to mock our efforts to "modify" what has already happened. Generalizing, we could agree that no past event is influenced or causally determined in any manner by ensuing conditions or events. Clearly PFF wants to discount the efficacy of our present actions which are intended either to bring about causally, or to avert future happenings: "there is nothing anyone can do about" the future. But does PFF want to broaden this, as he did with regard to bygone events, and deny that anything which happens now can affect things to come? If he stops short of this apparent rejection of all causality, he must give up thinking "of the future in the way we. . . think of the past". His most germane remarks go as follows:

> Determinism. . . is the theory that all events are rendered unavoidable by their causes.. . . [Some] distinguish this from fatalism by saying that, according to the fatalist, certain events are going to happen *no matter what*, or in other words, regardless of causes. But this is enormously contrived. It would be hard to find in the whole history of thought a single fatalist, on that conception of it.[10]

Perhaps so, historically speaking. But if he disallows "no matter what" versions of his theory, how can PFF maintain that future and past are causally symmetrical? Having thereby acknowledged present causes of things to come, should not PFF concede that what happens now may influence prior events?

I do not know of any fatalist who has explicitly taken that line. But it is noteworthy that one much-discussed rationale for fatalism, by PFF himself, appears to smuggle in a form of retro-efficacy. This is far from obvious, since PFF eschews all talk of causal relations between events. According to his official, putatively noncausal terminology, one "state of affairs" or "condition"

> is *sufficient* for, though logically unrelated to, the occurrence of some further condition *at the same or any other time*, [in case] the former cannot occur without the latter occurring also.... Another way of saying the same thing is that . . . one state of affairs *ensures* without logically entailing the occurrence of another. . . . Ingestion of cyanide, for instance, *ensures* death under certain familiar circumstances.[11]

Correspondingly:

> if the occurrence of any condition is *necessary* [or essential] for, but logically unrelated to, the occurrence of some other condition *at the same or any other time*, then the latter cannot occur without the former occurring also. . . .Oxygen, for instance, is *essential* to (though it does not by itself ensure) the maintenance of human life – though it is not logically impossible that we should live without it.[12]

One key "logical consequence" PFF extracts from these definitions of sufficient and necessary conditionship is that if *A* is sufficient for *B*, then *B* is necessary for *A*, and conversely – whether *A* occurs before, after, or simultaneously with *B*. He also makes it plain later that if *B* is necessary for *A*, and *B* fails to occur, then neither does *A*. Hence the 'nonoccurrence' of *B* suffices for the nonoccurrence of *A*. In particular, if *A* is someone's action, and *B* some "condition necessary for the occurrence of that act", then *B*'s nonoccurrence – prior to, during, or after the time for carrying out *A* – makes it impossible for the person to perform *A*.[13]

PFF's main example, like Aristotle's, is a forthcoming naval engagement; and his fatalistic thesis will be that it is no more within the power of anyone – including participants – whether a combat is going to occur tomorrow, than it is "whether or not a . . . battle has taken place" yesterday.[14]

Nobody would quarrel with this view of past battles.Yet PFF's reasoning in support of this half of his comparison seems uncharacteristically roundabout and forced. My hunch is that he wants a case in which there might appear to be room for backwards causality. His announced thesis is that it is not up to us now what has already happened or what will happen. So he will refute the suggestion that we might do something which is sufficient for the occurrence or nonoccurrence of a sea fight yesterday. Nevertheless he will have a pattern at hand, with a semblance of later-to-earlier sufficiency built into it, which he can exploit when he turns to the forthcoming battle. This will become clearer shortly. First here is PFF's 'past' example:

> ... I am about to open my morning newspaper ... conditions are such that only if there was a naval battle yesterday does the newspaper carry a certain kind (shape) of headline – *whereas if it carries a certain different sort (shape) of headline, this will ensure that there was no such battle.* ... I am about to perform one or the other of two acts, namely, [S] one of seeing a headline of the first kind, or [S'] one of seeing a headline of the second kind. ... If I perform act S, then *my doing such will ensure that there was a naval battle yesterday* ..., whereas if I perform S', then *my doing that will ensure that no such battle occurred.*[15]

PFF is right that it is not "up to me which sort of headline I shall read".[16] He explains that either the engagement has taken place or it has not; so either a necessary condition for act S, or else a necessary condition for act S' "is, or was, lacking".[17] All this is so elementary that we are apt to pay little attention to crucial phrases which postulate that my current actions – S and S' – suffice for ("ensure") the occurrence or nonoccurrence of a sea fight yesterday. Hence my emphasis.

I insinuated above that this brings in backwards causation. How does that square with PFF's deliberate avoidance of causal idioms? I am forced to read "ensure" causally by PFF's own attempts to illustrate his technical notions. How else can the swallowing of cyanide ensure death except by helping to cause death, given "familiar circumstances"? Does not an oxygen-rich environment contribute causally to "the maintenance of human life", and is it not "essential" inasmuch as scarcity of oxygen brings about cessation of human life?

More significantly, perhaps, PFF's illustrations of necessary and sufficient conditionship embody either simultaneous or forwards causation. Could anyone's death yesterday be "ensured by" the victim's ingurgitation of cyanide today? A future supply of oxygen does not seem "essential" to present human survival, in the way that current supplies are indispensable. An imminent shortage hardly ensures the ending of life in the straightforward manner of a shortage right now. A group of submarine trainees might learn with horror that their oxygen tanks will run out long before their vessel can reach the surface. They might then panic, and die almost immediately – although their oxygen lasts several more minutes. Was this catastrophe ensured – much less brought about – by future oxygen deprivation? No – the recruits' present fearful rioting constituted the "sufficient condition", or main causal factor, of this nonoccurrence of life, under the circumstances.

Thus I object to PFF's 'past battle and current headline' example because he has loaded it with these dubious backwards causal assuptions regarding sufficient and necessary conditionship. Such assumptions about ensuring past battles or nonbattles undermine his principal thesis that both past and future events events elude our control.

Retro-efficacy becomes more noticeable when PFF explains why tomorrow's sea fight is not up to us. Now his protagonist is not an unassuming newspaper reader, but someone in an ostensibly more influential position: the commander of one of the opposing Aristotelian flotillas. Again we only hear of necessary and sufficient conditions, not causal relationships. PFF tells us that if the admiral presently orders an attack upon the enemy fleet, that 'ordering' event, O, will suffice for a combat tomorrow, under prevailing circumstances; and his promulgation of a 'no attack' order, event O', will ensure that no fracas occurs.[18] Our philosophical problem, then, is whether it is within the admiral's power to issue whichever command he pleases. Thus far, incidentally, PFF seems to countenance forward causation – under the guise of sufficient conditionship – between the admiral's order-giving and later events. What else could sufficiency amount to here?

Anyway, the most significant dialectical moves come next, and no critic I have read seems aware that they introduce retro-efficacy. PFF remarks that on his definitions of sufficient and necessary conditionship, a battle tomorrow is necessary for event O (the attack order) today, since O suffices for the battle. Similarly, nonbelligerency tomorrow is necessary – essential – for O' (the 'no attack' order) today. If we wanted to reveal backwards causation quite starkly, we could add two further logical consequences of the earlier definitions: the nonoccurrence of an engagement tomorrow suffices for the nonoccurrence of O today, and a melee tomorrow suffices for the nonoccurrence of O' at this moment.

PFF could straightaway draw fatalistic conclusions about the admiral's control over today's orders and tomorrow's combat. Instead, perhaps to hide the backwards causal assumptions in his reasoning, he nimbly jumps semantical levels. He turns away from future happenings, and considers future-tensed propostions about them – specifically "the two propositions, 'A naval battle will occur tomorrow' and 'No naval battle will occur tomorrow' " – which he abbreviates as propositions Q and Q'.[19] Starting from the unassailable principle that either Q is true, or Q' is true, PFF argues:

If Q is true, then it is not within the admiral's power to do O' (for . . . then there is, or will be, lacking a condition essential for. . .doing O', the condition, namely, of there being no naval battle tomorrow).[20]

The same with event O in case proposition Q' is true. So it appears that either the admiral cannot order an attack, or else that he cannot order a retreat. Hence it is not up to him which he does, and he has no power over tomorrow's events, any more than over yesterday's.

PFF's remarks on the truth of propositions Q and Q' sound merely diversionary. What matters is not that proposition Q is true, but that "there is . . . lacking a condition essential for . . . doing O'". And what this deficit amounts to is that the enemy flotillas will fight tomorrow. Allegedly, the admiral requires "no naval battle tomorrow" in order to do act O' today (command a retreat). Equivalently a future clash suffices for his nonperformance of O' now. However we state his view, PFF seems to presuppose later-to-earlier efficacy.

Ultimately we shall look at another argument in which considerations of future-tensed truth and falsity do carry a dialectical burden. First we should grapple with backwards causation – or conditionship – in his naval anecdote. We should ask two unsubtle but inescapable questions: How could nonbelligerency tomorrow be needed, or essential today for the admiral to call a retreat? Conversely, how could an engagement which is going to occur, but has yet to begin, already suffice for his not ordering a pullback?

PFF amply illustrates "the ides of anything being essential for the accomplishment of something else".[21] All his examples look unmistakably causal, but none exhibit later-to-earlier relationships. For instance, in conformity with his general definition of 'essential', he says:

I cannot . . . live without oxygen, or swim five miles without ever having been in water, or read a given page of [Cyrillic] print without having learned Russian, or win a certain election without having been nominated.[22]

Replying to critics, PFF declares outright that essential conditions which postdate an occurrence are on a par with those which precede or accompany it. Unfortunately, his example of this parity takes negative and schematic form:

Whatever may be my abilities, it is no more within my power to exercise those abilities in the absence of some *past* condition necessary for doing what they enable me to do . . . than for the finest pianist to exercise his virtuosity in a room which contains no piano . . . . [For] the same reason . . . whatever may be my abilities, it is certainly *not within my power to exercise those abilities in the absence of some future condition necessary for doing what they enable me to do.*[23]

This does not answer our first question. PFF should not just assume that there might be "some future condition necessary for doing" actions like O and O'. He should give cases. He may take for granted our understanding of why present availability of oxygen is indispensable for the survival of organisms, and why they perish in its "absence". These are well-established empirical matters. Similarly if we interpret the 'swimming and water' example as a claim that people need some practice in water before they can begin to master any swimming technique. The reading example is equally uncontroversial. Only people who have received some training will decipher printed words at all reliably. Those who never receive coaching or practice usually prove to be incompetent at reading. Needless to say, all these examples of necessary conditionship are causal. The availability of oxygen, water, and language training contribute causally to one's survival, one's swimming, and one's reading. The 'election' example is trickier. In many societies it is a convention, or procedural rule, that the nominee who receives the most votes qualifies as the winner of an election. No matter now many votes a write-in candidate gets, if she or he was not nominated beforehand, she or he does not count as the victor. Here, nomination does not seem to contribute causally to victory, in the way that television advertising helps. But in a group which allows write-in candidates to win, nomination might help to produce victory.

With regard to the piano-playing example, it may be just another empirical fact that most musicians need a piano to produce certain notes – and that if they lack a piano, the sounds they get by tapping their fingers on a violin, or some other resonant surface, always turn out to be disappointingly substandard. Absence of a piano suffices, causally, for the nonoccurrence of piano-playing; and the presence of one would help bring the performance up to par. Alternatively, we could take a 'conventional' or 'definitional' stance, and declare that no sound-producing activity shall qualify as piano-playing unless the sounds issue from a piano.

Could there be 'future necessary condition' analogues of these cases? We have already tried, unsuccessfully, to imagine circumstances in which a future supply of oxygen is needed for life now, and an impending shortage could already have had lethal consequences. To swim now, you do not need future immersion or practice. Nor does your current reading performance require, or benefit from, subsequent learning exercises. My present endeavors to read Russian do not suffer because I am *going* to miss next week's lesson. The election example is unproblematical if we treat it causally. Suppose unnominated candidates are allowed to win if they receive the most votes. Now consider such a vote-getting candidate. How could her or his victory require subsequent nomination? *Ex hypothesi*, all victory required was the most votes. If we look at the situation conventionally, however, we can dream up a society which permits unnominated candidates to receive votes, but if one such aspirant has received a plurality, the rule is that he or she must then go through a nomination ceremony. Notice, however, that this makes official victory coincide with nomination. So nomination is necessary for electoral victory, but not subsequent to it. Moving on to the last example: How could it be essential to a virtuoso's present performance that a piano will also be available after he has finished? I suppose you could make it true by convention, or definition, that no musical display can count as piano-playing unless a piano remains on hand afterwards. But this fantasy does not help us understand how "there being no naval battle tomorrow" could be necessary for the admiral's ordering a retreat, or how a battle tomorrow might ensure the non-occurrence of a retreat order.

I have been forced by PFF's examples of necessary and sufficient conditionship to interpret these notions causally. No other reading makes sense, and we cannot simply take them as primitive, undefined terms. If we did, and meekly accepted as brute axioms whatever PFF says about the relationship of later to earlier events, it would be child's play for him to prove fatalism.

On the other hand, we should not dismiss his arguments for fatalism merely because they seem to depend upon hidden retro-causal notions. We ought to satisfy ourselves that later-to-earlier causation – or conditionship – really cannot be presumed in fatalistic reasoning. The argument we have been examining did not officially countenance backwards causation. But in writings not concerned with fatalism, PFF turns 180 degrees and upholds the possibility of retro-causation. In the first work I shall consider, PFF also explicitly equates causes with conditions. PFF and his distinguished coauthor assert that " '*A* causes *B*' means: *A* is sufficient for *B*", adding that "surely there is nothing in the causal relation, other than this relation of sufficiency".[24] They regard it as "merely a point of vocabulary" that people ordinarily refuse to call some condition a cause of other events if it occurs after those events.[25] Given the authors' assimilation of causation to atemporal sufficiency, they prove with ease that whatever reasons we have for supposing earlier events bring about later ones count equally in favor of later-to-earlier causation, and that any arguments against retro-efficacy also undermine our belief that future events may result from present happenings. Unfortunately they do not explain how something future – tomorrow's battle or nonbattle – could suffice, presumably in a causal manner, to limit our admiral's present behavior. As far as I can tell, these authors would first appeal to the premises of our example, which make alternative acts of the commander sufficient for a battle or no battle, respectively. Then they would question-beggingly invoke their definitions, which make tomorrow's battle in turn sufficient for the admiral's not ordering a pullback, and nonbelligerency sufficent for his not ordering an attack. But why should we accept these strange causal assumptions?

For help with this, I will look at two other works by PFF on retro-efficacy.[26] In the terminology of both, we would say that a naval combat tomorrow "postvents" the admiral from calling a retreat now. Luckily, postvention gets full coverage. One of PFF's illustrations goes as follows: it is a necessary condition for a man's taking breakfast with his wife at some regular hour – say 8 A.M. – that he wakes up a few minutes beforehand. Consequently, explains PFF:

> By breakfasting with [him] ... at the usual time [8 A.M.] the wife guarantees, in the best possible way, that he has not overslept [i.e., that he has awakened at the earlier time]. The wife, accordingly, can properly claim to have postvented his oversleeping by breakfasting with him.[27]

In these writings, PFF modifies his position somewhat. For he denies that postventing – and I suppose 'guaranteeing' – amounts to causing. In fact he discerns "something metaphysically absurd... in supposing that efficient causes might work 'backward' ".[28] This absurdity he exhibits in another imaginary situation: A politician dies. Under the circumstances which hold before and after this sad event, the actions of two other people are sufficient conditions for the death. Beforehand, a rival of the politician shot him. And after the death, a friend of the late leader attended funeral ceremonies for the deceased. PFF askes: Why not say that both men's actions brought about the politician's demise, and that the funeral guest and the gun-slinging opponent are equally to blame for it? PFF's answer is that

> no cause exerts any power over the past. ... [A] man who shoots another *acts upon* him. ... The man who merely attends the funeral ... does not act upon him who is already dead. ... For anything to be a cause it must act upon something and, as a matter of fact – indeed of metaphysical necessity – nothing past can be acted upon by anything.[29]

I suspect that it only labels, but does not solve a problem like this when one speaks of "metaphysical necessity" or absurdity. We surely want to learn just why it is cosmically *verboten* for causes to act upon something bygone. So we may gain insight if we probe further. A minor point worth noticing is that after the politician's demise, we cannot act upon him in any manner which amounts to 'causing his death'. You cannot assassinate a cadaver. More significantly, even if the politician were somehow alive during funeral services for him, nothing you do to him then would rank as 'causing him to have died at an earlier time'. Thus even if a person or thing 'from out of the past' somehow becomes available to be "acted upon", the outcome will never be a prior event, such as the person's antecedent death.

Again we should inquire: Why not? In answer, I shall try out what sounds to me like a quasi-empirical suggestion of Dummett's,[30] recently elaborated by Sarah Waterlow.[31] We should focus on relatively prolonged events, or continuing processes, which begin and then endure or evolve for a while. As Waterlow remarks, our present actions frequently set off processes which persist into the future – but none which unfold backwards to an earlier period. Onward-moving events may terminate, or fail to terminate, in some end state that you hoped for. They may, in their turn, initiate further processes that you desired, or they may prove barren. The interesting point is that this is our only available method of achieving the end states and further processes in question. Consequently, if you hanker for end states and enduring events that belong to an earlier time, you are out of luck – since no processes appear to unfold in that direction.

To illustrate these very general pronouncements, I shall reintroduce the terrorist bomber. This time he is only experimenting with various missiles and their component parts. With a lighted match he ignites a mesh fuse, and it burns all the way through – as he hoped it would. In a further test, he attaches the end of a similar fuse to a bottle of gasoline. After he lights this fuse, it sputters away steadily until flames reach the neck of the bottle. Suddenly there is a quite different, loud and violent process. The *aperçu* I borrow from Dummett and Waterlow is that events like steady combustion, or the sequence, combustion followed by the gasoline exploding, only continue forward in time. I regard this as a semi-empirical fact because it is not part of what we mean by 'process' that such events must unfold from earlier to later. On the other hand, can we imagine the opposite? What would it be like for combustion, or any other sort of persisting event, to begin now but end before now?

Rather than plunge into that mystery, we should go on to another of PFF's reasons for distinguishing what he calls "postvention" from genuine backwards causation. He considers situations where we act for a purpose – where our deed is intended to be a means of attaining, or blocking, some result or circumstance. PFF acknowedges that nobody would ever take 'postcautions' or 'postventive measures' in order to stop earlier catastrophes from having already occurred. We have no procedures, reliable or hit-or-miss, for bringing about or averting past events. Regarding his own example of postvention,

PFF declares it "plainly untrue that the wife breakfasted with her husband as a means of getting him" not to have overslept; even if it is one of her chief "goals or purposes" to have him awaken before the scheduled breakfast time, her action of breakfasting with him is not a "suitable or. . . intelligible means" of pursuing her objective.[32] I should add that in cases where you are uncertain whether some event has happened, and do something now in order to find out whether it did, you scarcely make the past event to have, or not to have occurred. For instance, a skier takes a wrenching spill. Her ankle hurts. If she broke it, it will not support her full weight. So she pulls herself up and manages to stand. This guarantees, in an epistemic rather than causal sense, that her ankle did not break a few minutes before. She did not postvent it from having snapped.

We have nearly eliminated the kind of retro-efficacy that seemed to underlie PFF's story of the admiral. But according to PFF, there is one limited but unambiguous situation where a person's current action really causes, and serves as a means of producing, earlier events. PFF singles out

> purposive actions, the ends or goals of which seem actually to precede those actions. For example, if it is one's purpose to flex a certain arm muscle, the only way he can accomplish this, normally, is by moving his arm. The arm is *caused* to move by the motion of the muscle, but the arm is moved in order to move the muscle. . . . The motion of the muscle does not follow the motion of the arm . . . [but] precedes it slightly. Or consider nerve impulses . . . [which help] cause . . . the motion of a man's arm when he moves it. . . . If a man learns of these nerve impulses . . . and then has . . . to produce one . . . for some experiment . . . he can do so only by moving his arm. . . . Moving his arm . . . is the means to . . . the production of a certain nerve impulse – which actually precedes that purposive action. . . . [This] is the only kind of case I can imagine of changing the past *at will*, of having some event in the past literally within one's voluntary control.[33]

Should we allow these as isolated exceptions? I think we need not. As I have argued elsewhere,[34] we should not confuse a person's action of "moving his arm", and the strictly peripheral event, "the motion of the arm". Why distinguish them? Because our limbs often gyrate around, or remain motionless, when it is clear that we are not moving them – or holding them still. Examples are when a physician tests our patella response; when our limbs flail during our sleep; when our bodies shiver uncontrollably; when our joints suddenly stiffen from 'charley horse'; and when we undergo convulsions. None of these limb motions are instances of the person moving the affected part of her or his body.

What else, over and above peripheral "motion of the arm", is needed for the action of moving one's arm? My proposal is to include, as subevents making up the action, the right kind of "nerve impulses" and muscular contractions. Arm-waving is an action that I perform. Crucial things must go on in me besides peripheral motion. If there are no nerve impulses to initiate reactions of tendons in my neck and shoulders, and if there are no feedback nerve impulses to regulate the continuing peripheral motion, the upshot probably will be unsatisfactory. That is, the resulting peripheral motion will not qualify as the kind of gesture I intended. So my proposal is to make all these indispensable neural and muscular goings-on integral elements of the larger event which is my action of arm-waving. I would also add crucial psychic goings-on. For instance, I am conscious – that is, awake – and I am conscious *of* various things: the initial position of my arm, its trajectory as it moves; the proximity of my friend Max; the luncheon engagement we have; and of the social convention of greeting someone by waving an arm. It is also relevant that I desire to wave my arm, to hail Max, and so on. Perhaps emotions, like joy at seeing Max for the first time in months, figures as part of my action – if it is a joyous wave. These mental goings-on or states are elements of my performance because they are the cognitive, conative, and affective attitudes on which I act. If you are a materialistically-inclined metaphysician, you may want to identify the psychological subevents on my list with some of the initiating neural subevents I mentioned. That is, you might hypothesize that, as a matter of contingent fact, the beliefs, desires and emotions on which I am acting are the very same particular subevents or states as certain brain processes. If you are a metaphysical dualist, of course you will count mental and neural goings-on as altogether disparate components of my deed.

One dividend of this account of action is that we can say what other subevents it includes, over and above agitation of one's body. Probably the sequence of neural and muscular happenings which occurs when we are moving a limb is easy to distinguish from what occurs when a physician tests our patella response, or we flail about in our sleep, or undergo shivering, stiffening of the joints, or general convulsions. Surely we can make the contrast by reference to mental subevents. The thrashing sleeper is neither conscious, nor conscious of the motions of his or her limbs. The other people may be awake, and realize how their limbs are moving, but have no desire for such limb motions. Also, if my desires changed, and I did not want to wave, my arm would remain still. But the patella response, the shivering, stiffening, and convulsions occur even when the person wants them to stop.

For our purposes, the main benefit of this component-event analysis of action is that we can reject backwards causality. PFF's bothersome "motion of the muscle" and "nerve impulses" are integral parts of my arm-waving action. They do not precede it, even "slightly". It begins with the initiating nerve impulses and the onset of my desire or decision to wave now. Peripheral motion does result from these neural and psychological events, somewhat later. But the global event which is my arm-waving action has already begun. No causal relations can hold between the global action-event and its neural or muscular event-components. They are parts of it; so they do not forwardly cause it, and it does not backwardly cause them to have occurred. All we need acknowledge is forwards causation among event-components which make up the action. So I do not have "some event in the past literally within" my "voluntary control". At most, some nerve impulses and muscular contractions, which are contemporaneous with the beginning of my arm waving action, may fall under my control. I get them to start now by starting to swing my arm now. Slightly after my action begins, there will be peripheral motion of my arm. But peripheral motion is not needed as a cause of the antecedent neural and muscular goings-on which engendered it. If you wonder, "What then did produce those impulses and contractions?", you will find plenty of forward-operating causes. Mind-brain dualists may suppose that my beliefs, desires and emotions are nonphysical events which somehow jogged my neuromuscular system into motion. If you are a materialist, you will look for earlier physical causes, within or outside my brain, for the nerve impulses which you equate with the cognitive, conative, and affective beginnings of my arm-waving action.

We may neglect such details. Backwards causation came into our discussion of fatalism because PFF's argument made a future naval battle sufficient for the admiral's not ordering a retreat now. Finally, however, we may dismiss the threat of later-to-earlier sufficiency. There could be no such analogue of familiar, presumably causal, forwards sufficiency. But this result does not yet entitle us to shrug off fatalism. PFF has another challenging rationale for that doctrine which exploits past-future symmetries without appealing to retro-efficacy, or even to necessary and sufficient conditions.

This new argument hinges upon considerations of the truth of future-tensed propostitions – which surfaced briefly but inconsequentially in the sea-fight case. Readers know, of course, that these considerations figure in the most ancient rationales for fatalism.[35] For his part, PFF wants to convince us again that future events are no more within our power than bygone happenings. But now his key premise is semantical. He says:

> there is a body of truth concerning what the future holds, just as there is a body of truth concerning what is contained in the past, whether or not it is known to any man or even

to God; and hence everything asserted in that body of truth will assuredly happen, . . . precisely as it is described therein.[36]

To illustrate, PFF replaces our Aristotelean naval commander by an obscure high-school teacher named Osmo. Instead of having his present range of action restricted by future naval battles, Osmo runs afoul of a mysterious book which is filled with true statements about his past and – apparently – his future career as well. The statements are not in either the past or future tense, but they convey equivalent information by specifying, in the 'tenseless present' what happens on various dates. Most interestingly, there are statements recounting how and when Osmo will perish. He is skeptical. But just before he dies in a frantic attempt to escape his doom, Osmo acknowledges that fatalism is correct. PFF explains Osmo's verdict, once more in fundamentally semantical terms. He tells us that

Osmo became a fatalist because there existed a set of true statements about the details of his life, both past and future . . . That is the whole of it . . .

. . . Even . . . when Osmo tried so desperately to save himself from the destruction described in the book, his effort was totally in vain . . . [since] what was said there was true. No power on earth can render false a statement that is true. It has never been done, and never will be.

. . . [The fact that Osmo] came to know what those statements were and to believe them has nothing to do with the validity [of his fatalism][37]

Here, apparently, truth does not set us free. PFF is explicit about how the truth of statements curtails our power to shape events. He reiterates:

No one has ever rendered false a statement that was true, however hard some men have tried. When [we attempt] . . . to avoid a given calamity, and the thing happens anyway, at just the moment and in just the way it was going to happen, we have reason to doubt that it could have been avoided. . . .We can say that *if* more strenuous precautions had been taken, the event would not have happened. But . . . we must add [that] they were not taken, and hence . . . they were not going to be taken, – and we have . . . again a true statement of the kind that no man has ever had the slightest degree of success in rendering false. . . .

. . . The fatalist argument . . . has to do with unavoidability . . . with human abilities. . . . No man is able to avoid what is truly described . . . in any statement, nor to bring about what is . . . falsely described. . . . It has never been done, and it never will be.[38]

My first reservation about this goes back to our central theme: the alleged parity of bygone and future happenings. Recall why we think ourselves impotent vis-à-vis what has already taken place. Do we adopt this stance because "there is a body of truth . . . concerning what is contained in the past"? That sounds beside the point. A more natural, and ultimately reasonable, answer would be that we are helpless because earlier events have already run their course, and because nothing that goes on now can possibly influence them. For instance, Osmo bicycled yesterday to the school where he teaches, and while doing this got caught in a sudden downpour. Because these incidents are over and done with, it is too late for Osmo to have taken the bus and stayed dry. Also, because they have already occurred it is true to report these events in the past tense. It would be a strange inversion if we said: 'Because our past-tense report is true, nothing can be done to avert Osmo's cycling and drenching'. The 'because' I am harping on is elusive. It cannot mark either a causal relation or a logical entailment between event and report. Perhaps it conveys a very rustic kind of ontological primacy, which might be illustrated as follows. We can easily imagine a universe containing no statements but lots of events – with physical objects, some changing and others persisting more or less the same. But can we intelligibly imagine or describe a universe containing no events, but only statements or propositions – which are, moreover, either true or false in the way that 'A naval battle will occur tomorrow' and 'Osmo loses his life in a plane crash on 5 June 1981', are either true or false? Of what would they be true or false?

This line of inquiry may give sense to what I mean when I suggest that events have an ontological head start over true event-propositions. At any rate it should be clear that the fatalistic argument I just quoted fails to maintain past-future parity. We have far weightier reasons for saying that bygone events elude our control, besides PFF's "body of truth . . . concerning . . . the past".

But should a fatalist really want to uphold such a view of why we cannot influence future events? Of course he should not renounce a "body of truth" thesis regarding the future merely because it fails to explain our helplessness toward bygone events. What matters for him is to prove that future happenings are "unavoidable". The reasons may differ from those which account for our inability to shape the past. So we will forget 'parity' claims, and ask if "a body of truth concerning what the future holds" could make the predicted events inescapable. Straightaway the fatalist must refute our intuitions of ontological primacy, that just as past events do not elude our control because of true past-tense statements recording them, neither do future events because they are truly prophesied. Maybe these intuitions are too vague to prove anything. But then we can reintroduce the "no matter what" problem with which we confronted PFF at the outset. We say to "body of truth" fatalists: Do you mean that Osmo's death will occur only because the statement, 'Osmo loses his life in a plane crash on 5 June 1981' happens to true? Are there not reasons or causes for his boarding tht flight, and causes of the disaster? PFF's reply is that Osmo did not . . . believe that certain things were going to happen to him *no matter what*. . . . if anyone believes that a given event is going to happen, he does not doubt that those things necessary for its occurrence are going to happen too.[39]

This reply has an awkward consequence. A future event will be overdetermined. It will occur because there is a true statement forecasting it, *and* because of "those things necessary for its occurrence". Presumably it would be a sufficient condition of some kind if we only had the true prophesy, but none of the other "things" or causes. So if a fatalist maintains that "those things . . . are going to happen too", he must assume a miraculous coincidence, or preestablished harmony, between true predictions of each event to come, and the occurrence of whatever it takes to bring about the event. But why should we always have both?

Even if we swallow overdetermination and this miraculous match between causes and true forecasts, there is a hitch. The fatalist's 'inevitability' conclusion does not seem to follow. Our premise, that Osmo's crash was overdetermined by the true statement and by whatever caused his crash, hardly entails that this mishap could not have been averted. Nor is that outlook make plausible. Here is why. We can assume that PFF is not merely offering us the much-disputed 'hard determinist' argument that "events are rendered unavoidable by their causes". So Osmo's demise in 1981 is not inescapable by virtue of having some cause or other. Now the only other factor mentioned by PFF is the true statement. But how does it make the crash something we cannot prevent?

"Body of truth" fatalism seems to beg the main question under debate, namely: Is it ever up to us what will happen? Notice how we are first given categorical, nonmodal premises, which I restate more or less in PFF's words:

> everything asserted in the body of truth *will* happen;
> no power on earth *has* rendered false, or ever *will* render false, a statement that is true;
> when we attempt to avoid a calamity that *is* going to happen, the thing *happens* anyway, in just the way it *was* going to happen;
> more strenuous precautions *were* not taken;
> if it is true to state that more strenuous precautions *will* not be taken, then we have a true statement of the kind that no man *has ever had* the slightest degree of success in rendering false.

Suddenly we are astonished to hear various modal conclusions:

everything asserted in the body of truth *must* inevitably happen;
no power on earth *can* render false a statement that is true;
we have reason to doubt that the calamity *could* have been avoided;
nobody *could* have taken more strenuous precautions;
if the statements, 'More strenuous precautions *are* not going to be taken', and 'The calamity *is* going to occur', truly describe future events, then no one is *able* (has the ability) to avoid these events.

Have truth-fatalists covertly added the question-begging premise that whatever is going to happen is going to happen *inevitably*? Unless they have assumed this, they are only entitled to a far more modest type of conclusion: that events corresponding to the true statements *will* take place, and that effective countermeasures *will* not be deployed.

The dialectical situation is metaphysically befuddling, of course. We can not point to an actual case, and say: 'A disaster was going to occur; that bridge over there was going to fall. The statement, "This bridge will give way", was a true statement. But fortunately the inspectors knew how to detect such dangers ahead of time; and our engineers, together with their work crews, had the skills, implements, and materials they needed to repair the structure. Everyone did what was required to avert the impending catastrophe, and it did not occur'. PFF is quite right: there will never be such a case. If the proposition, 'This bridge will collapse' is true, then it is logically impossible for us to produce an instance where we successfully display our methods and means of heading off the predicted mishap. We can find cases which are alike in many vital respects – where a bridge was equally flawed, and people used the same techniques and machines to prevent its collapse. But then it would not be *true* that the bridge was going to crumble. So our position is uncomfortable. However, a truth-fatalist's may be worse. For he merely assumes his 'inevitability' thesis, instead of demonstrating it by argument, if he infers that we cannot forestall a truly-described calamity. He ought to explain why we cannot. Does he think us deficient in strength, skill, intelligence, information, instruments, authority or something else required to prevent disasters? Does a true proposition mysteriously neutralize the abilities and so forth that we have exercised on other occasions? How does it operate? Like paralysis or fatigue?

Truth-fatalists might argue that our abilities and so on will be undermined by another true proposition: "We are not going to exercise our abilities". But that reasoning is circular. It presupposes what is at issue – that if an event, such as the nonexercise of our abilites, is truly predicted, then the event must inevitably occur. If this is where truth-fatalism leads, we have no more reason to accept it than the argument from necessary and sufficient conditions.

I conclude that we need not think of the future in the way we think of the past. We can believe that what will occur often depends upon what happens now and what we do now. This belief does not commit us to supposing that bygone events might also be molded by present goings-on. Nor does our denial that any current happenings affect the past require of us a similar attitude toward what is to be. I have chiefly endeavored to rebut the fatalistic contention that future events are no more 'up to us' than past occurrences. So I have perhaps not built an airtight case against backwards causation. Nevertheless we have noticed repeatedly how obscure the hypothesis seems to be. None of the putative examples were coherent, and we found ourselves unable to devise analogues for the orthodox instances of simultaneous and forwards efficacy. By contrast with the rich detail available to us about what goes on when present events determine the sequel, we have no understanding how they might instead cause – or be sufficient or necessary for – things to have already taken place. So while I may not have demonstrated the impossibility of backwards causation, I have shown that its proponents owe us an account of what it might be like.[40]

The University of Illinois at Chicago Circle

NOTES

[1] D. F. Pears, 'Time, Truth, and Inference', *Proceedings of the Aristotelian Society* **51** (1950–1951), 1–24.

[2] M. A. E. Dummett, with A. G. N. Flew, 'Can an Effect Precede Its Cause?', *Proceeding of the Aristotelian Society* (supplement) **28** (1954), 27–62.

[3] R. M. Chisholm and R. Taylor, 'Making Things to Have Happened', *Analysis* **20** (1960) 73–78.

[4] Gilbert Ryle, 'It Was to Be', in *Dilemmas*, Cambridge University Press, Cambridge, 1954.

[5] R. D. Bradley, 'Must the Future Be What It Is Going to Be?', *Mind* **68** (1959), 193–208. Reprinted in Richard M. Gale (ed.), *The Philosophy of Time: A Collection of Essays,* Anchor-Doubleday, Garden City, N.Y., 1967, pp. 232–251.

[6] Richard Taylor, 'Fatalism', *Philosophical Review* 71 (1962), 56–66. Reprinted in Gale, *op. cit.*, pp. 219–231. The page numbers here refer to the Gale volume.

[7] Richard Taylor, *Metaphysics*, 2nd ed., Prentice-Hall, Englewood Cliffs, N.J., 1974, p. 60f. See also his 'Fatalism', p. 221; Bertrand Russell, *Mysticism and Logic*, George Allen and Unwin, London, 1918, pp. 201ff.; A. J. Ayer, *The Problem of Knowledge*,

Penguin, Harmondsworth, 1956, pp. 16 – 175; and Steven M. Cahn, *Fate, Logic and Time,* Yale University Press, New Haven, Conn., 1967, p. 9.
8 Taylor, *Metaphysics,* p. 67f.
9 *Ibid*., p. 65f.; see Cahn, *op.cit.*, pp. 8ff., 15, 19, 23.
10 Taylor, *Metaphysics*, p. 59; see also p. 65 and Cahn, *op. cit.,* pp. 16 – 23.
11 Taylor, 'Fatalism', p. 223, some emphasis added.
12 *Ibid.*
13 *Ibid.,* p.223f.
14 *Ibid.,* p. 225.
15 *Ibid.,* p. 224, emphasis added; see also Cahn, *op. cit.,* pp. 32–44.
16 Taylor, 'Fatalism', p. 224f.
17 *Ibid.,* p. 225.
18 *Ibid.,* pp. 223, 226.
19 *Ibid.,* p. 226.
20 *Ibid.,* p. 227.
21 *Ibid.,* p. 224.
22 *Ibid.*, see Cahn, *op. cit*., p. 92, and Cahn, 'Fatalistic Arguments', *Journal of Philosophy* **61** (1964), 295–305, p. 297.
23 Taylor, 'A Note on Fatalism', *Philosophical Review* **72** (1963), 497 – 499, p. 498f., emphasis added; see Cahn, 'Fatalistic Arguments', p. 299f.; and *Fate, Logic and Time,* pp. 92–95.
24 Chisholm and Taylor, *op. cit.,* p. 73f.
25 *Ibid.*
26 Taylor, 'Prevention, Postvention and the Will', in Keith Lehrer (ed.), *Freedom and Determinism*, Random House, New York, 1966, pp. 65–85; *Action and Purpose*, Prentice-Hall, Englewood Cliffs, N. J., 1966, pp. 31–39, 188–195.
27 Taylor, 'Prevention', p. 70; see *Action*, p. 189ff., and also A. J. Ayer, 'Fatalism', in *The Concept of a Person and Other Essays,* Macmillan, London, 1963, pp. 253–268, p. 239.
28 Taylor, *Action,* p.33.
29 *Ibid.,* p. 34f.
30 Dummett, 'Can an Effect Precede Its Cause?', p. 31.
31 Sarah Waterlow, 'Backwards Causation and Continuing', *Mind* **83** (1974), 372–387.
32 Taylor, *Action,* p.193.
33 *Ibid.,* p. 194f.
34 Irving Thalberg, 'Free Will and Chisholm's Varieties of Casusation', *Idealistic Studies* **1** (1971), 149–159, p. 152ff.; 'Constituents and Causes of Emotion and Action', *Philosophical Quarterly* **13** (1973), 1–14, pp. 8–12; *Perception, Emotion and Action,* Blackwell, Oxford, 1977, pp. 66–71.
35 See J. L. Ackrill, *Aristotle's* Categoriae *and* De Interpretatione, Oxford University Press, London, 1963, pp. 53ff., 132–142.
36 Taylor, *Metaphysics,* p. 60.
37 *Ibid.,* p. 65ff.; see also "Diodorus Cronus"(i.e., Richard Taylor and Steven M. Cahn), 'Time, Truth and Ability', *Analysis* **25** (1965), 137–141, p. 139ff.
38 Taylor, *Metaphysics*, p. 69f.
39 *Ibid.,* p. 65.
40 I read a version of this paper at Illinois State University in December 1976, and

thank listeners for their comments. My colleague Daniel Berger gave me a number of valuable criticisms. Peter van Inwagen offered me detailed reactions to the penultimate draft, and kindly sent me two interesting manuscripts of his own on fatalism.

MARGERY BEDFORD NAYLOR

FATALISM AND TIMELESS TRUTH

One of Professor Richard Taylor's greatest contributions in philosophy is his work on fatalism. He has repeatedly attempted to make the strongest possible case for fatalism,[1] sometimes hoping to vindicate and other times to refute it once and for all. We might find his arguments compelling if we were not so certain that fatalism must be false. It is much easier, however, to see that something must be wrong with them than it is to say what is wrong with them. Yet we are obliged to refute each one or else admit that our denial of fatalism is unjustified. The purpose of this paper is to refute the argument which seems to underlie Taylor's most recent defense of fatalism in the second edition of *Metaphysics.*[2]

Taylor presents this defense using the story of Osmo, which (much condensed) is this. Osmo finds his complete biography – book containing true propositions about his future as well as his past. He concludes that every event of his life is one which occurs 'unavoidably'. This case can be extended to justify the universal and perfectly general thesis that *whatever* occurs occurs unavoidably.

In this paper I shall state in Part I what I take to be the argument the premises of which are the principles the story of Osmo seems to rest on and the conclusion of which is fatalism. Taylor has not owned these explicitly, although I believe he should be willing to do so. In the second and third parts I shall try to show that one of these premise makes the argument question-begging when it is used to prove fatalism, and that that premise is false.

I

Fatalism is the thesis that any particular event which occurs at some particular time is such that it is impossible for the occurrence of that event at that time to be avoided either before or during or after that time. As Taylor puts it, "Whatever happens is and always was unavoidable".[3] We shall state fatalism as (F).

(F) For any event, E, which occurs at some particular time, t, it is logically impossible for there to be any time when the occurrence of E at t is avoidable.

Peter van Inwagen (ed.), Time and Cause, 49–65.
Copyright © 1980 *by D. Reidel Publishing Company.*

In (F) '*E*' is replaceable by the name or description of an event-token rather than an event-type, since the event whose occurrence is supposed to be unavoidable is a particular event rather than a kind of event. In addition, '*t*' is replaceable by the specification of a particular time, which might be an instant (if there are instantaneous events) or an interval (if *E* has some definite duration). This time might be past, present, or future. The significance of 'logically impossible' in (F) is that according to (F) a contradiction follows from the conjunction of the proposition that *E* occurs at *t* and the proposition that there is at least one time when the occurrence of *E* at *t* is avoidable. If *E*'s occurring at *t* is 'unavoidable', then its not occurring at *t* is impossible. No one and no thing can make it the case that *E* does not occur at *t*.

The argument for fatalism to be stated here is one we can accept (as Taylor points out) without subscribing either to the deterministic thesis that "all events are rendered unavoidable by their causes",[4] or to the thesis that an omniscient or an omnipotent being makes whatever occurs occur unavoidably.[5] Nor must we accept any particular account of what it is to have the ability to prevent the occurrence of a particular event at a particular time or what it is to 'render' a proposition false. To accept the argument we need only grant three apparently innocent premises, to be called here, respectively, the 'universal fixity' thesis (U), the 'eternal fixity' thesis (E), and the 'sufficiency' thesis (S). Here, then, is our statement of the argument for fatalism.

(U) For any event, *E*, which occurs at some particular time, *t*, there is a true proposition that that event occurs at that time.

(E) For any true proposition that a given event, *E*, occurs at some particular time, *t*, it is logically impossible for there to be a time when that proposition is not true.

(S) If (U) and (E), then (F).

Therefore:

(F) For any event, *E*, which occurs at some particular time, *t*, it is logically impossible for there to be a time when the occurrence of *E* at *t* is avoidable.

In the premises '*E*' is replaceable by the name or description of an event-token and not an event-type. Something must be said about the term 'proposition'. It does not mean 'sentence', but rather whatever is expressed by a meaningful declarative sentence. Propositions – not sentences – are either true or false. Sentences are 'strings' of words, and they are not, strictly speaking, 'bearers of truth-value'. The same sentence may express different propositions in different contexts; for example, a sentence containing at least one indexical term (such as 'here', 'now', or 'I') expresses a different proposition each time it is uttered in a different context. It is also true that the same proposition might be expressed by different sentences, since differences in wording do not always entail differences in what is said.

The universal fixity thesis, (U) asserts that for any event (past, present, or future) which occurs at some particular time, the proposition that that event occurs at that time is true. Without (U) the remaining premises, (E) and a suitably amended (S), entail only the very weak version of fatalism, (WF).

(WF) For any event, *E*, which occurs at a given time, *t*, such that the proposition that *E* occurs at *t* is true, it is logically impossible for there to be a time when the occurrence of *E* at *t* is avoidable.

This is weak because if there were some events which occur but about the occurrence of which there are no true propositions, then there would be some events which occur whose occurrence is not unavoidable. Thus without (U) the argument could not establish that *everything* which occurs does so unavoidably.

The eternal fixity thesis, (E), asserts that any true proposition expressed by a sentence of the form '*E* occurs at *t*' is 'eternally true'. What (E) does *not* assert is that that true proposition is itself logically necessary, but rather that since the proposition is true it cannot *also* be false, and it cannot *also* fail to be true.[6] What (E) does assert is that the truth of the proposition that *E* occurs at *t* is fixed for all time even before *t*, not just forever after *t*. If the argument for fatalism lacked (E), (U) and (S) – suitably modified – would not entail that there is any time when the occurrence of a given event at a given time is unavoidable, so they would not entail fatalism.

The argument for fatalism is valid. We shall not question (S), which seems uncontroversial. Much of interest might be said about (U). Although I do not discuss (U) at all here, the arguments in Part II show that its verbs cannot be tenseless, and those is Part III show that (U) must be false for propositions about future events. The remainder of the paper is a discussion of (E). In Part II I attempt to refute one defense of (E), and in Part III to show that it begs the question and that it is false. If this is right, then fatalism must be false.

II

One might argue for (E) by appealing to a thesis defended by R. D. Bradley in 'Must the Future Be What It Is Going to Be?',[7] although Bradley does not use it to support fatalism. His thesis is that " . . . it just does not make sense to ask at what particular time a statement is true or false".[8] If this is true, then it must be meaningless to suppose that there is a time when a proposition which is now true was not yet true, for example. Thus, if Bradley is right, then (E) must be true if it is construed as (M).

(M) For any true proposition that a given event-particular, E occurs at a particular time, t, it is meaningless to suppose that there is a time when that proposition is not true.

I believe, however, that Bradley's claim is false, and that it *does* make sense to ask when a proposition is true. The next task here is to refute each of the three most important arguments Bradley gives for his thesis. Then, using the grounds for objecting to this thesis as a basis, we must construct a theory of what sense there is in asking when a given proposition is true.

Bradley's first major argument is this one.

(1) It does not make sense to say that p is 'true now' just because it makes sense to say that p is 'true always'. This is because 'p is true always' does not entail 'p is true now'' The latter is "ordinarily taken to mean 'true at this time but not at some other time' ", [9] and so it cannot be entailed by 'p is true always'.

This argument is based on the false premise that 'p is true now' is usually taken to *mean* 'p is true at this time but not at some other time'. The latter might be 'suggested' by (or be a Gricean 'implicature' of) 'p is true now', but 'p is true now' does not entail 'there is a time when p is false' or 'there is a time when p is not true'. The proposition that two plus two equals four is true now (since it is always true) and there is no time when it is not true, for example.

(2) Even if ordinary usage allows us to say that a '*grammatical* statement' is true at one time and false at another, it does not follow that it makes sense to ask when a proposition is true.[10]

The only reason it makes sense to ask when a grammatical statement is true is that it contains indexicals which may be specified in different ways in different contexts. Sometimes what it expresses is a true proposition, and sometimes what it expresses is a false one. But each of the propositions it expresses is true 'once and for all' if true or false 'once and for all' if false.

Bradley is right in pointing out that it does not *follow* that if ordinary usage allows grammatical statements to vary in truth-value it must therefore make sense to ask when a proposition is true. But this does not show that in ordinary usage any proposition must be true once and for all if true or false once and for all if false. If in ordinary usage even a proposition may vary in truth-value, then it does make sense to ask when a proposition is true.

Apparently ordinary usage does allow such variations. That this is so can be inferred from the assumption that Quine's proposal to reduce tensed sentences to tenseless ones which are "timelessly true" has a point. What Quine suggests (and Bradley endorses this proposal) is that

> . . . logical analysis is facilitated by requiring that each statement be true once and for all or false once and for all, independently of time. This can be effected by rendering verbs tenseless and then resorting to explicit chronological descriptions when need arises for distinctions of time. The sentence 'The Nazis will annex Bohemia', uttered as true on May 9, 1936, corresponds to the statement 'The Nazis annex [tenseless] Bohemia after May 9, 1936'; and this statement is true once and for all regardless of date of utterance. The shorter statement 'The Nazis annex Bohemia', construed tenselessly, affirms merely that there is at least one date, past, present, or future, on which such annexation takes place; and this statement again is true once and for all.[11]

It would make no sense to impose such a requirement, however, if 'The Nazis annex Bohemia' were not ordinarily understood as expressing a proposition which is not (if true) true independently of time. As Bradley himself points out, the sentence, 'The Nazis will annex Bohemia' is such that "According to ordinary usage, this assertion was true before 1939 but is now [i.e., after 1939] false".[12]

It might seem, however, that what Taylor calls the "journalistic present tense" (which – not incidentally – he uses in telling the story of Osmo's conversion to fatalism) is available from ordinary English to state tenseless sentences. We can imagine, for example, Lowell Thomas saying ominously 'The Nazis annex Bohemia' in a newsreel shown repeatedly in movie theater. The verb in this sentence is clearly not in the simple present tense, or else the proposition it expresses would be false if the Nazis did not annex Bohemia at the time when the film was shown. A sentence in the journalistic present tense is not tenseless, however; it is in a peculiar form of the present tense (what might be called the 'pseudo-present' tense). The journalistic present tense allows us to speak of an event – no matter whether that event is past, present, or future relative to the time the sentence is uttered – *as if* the event were occurring now, i. e., *as if* the simple or continuing present tense were appropriate. The use of the journalistic present tense is not to eliminate, but only to disregard, the appropriate tense.

It is not clear that adopting Quine's proposal will insure that it makes no sense to ask when a proposition is true. If tensed sentences express proposition which are not true (if true) or false (if false) independently of time, then if tensed sentences can be translated *without any loss of meaning* into Quine's tenseless ones, then tenseless ones cannot be true (if true) or false (if false) independently of time either.

Tensed sentences cannot, however, be translated without any loss of meaning into tenseless ones.[13] We can see that this is so by considering the follwoing tensed sentences.

(a) The Nazis annexed [simple past tense] Bohemia.
(b) The Nazis annex [simple present tense] Bohemia.
(c) The Nazis will annex [simple future tense] Bohemia.

These three sentences express three irreducibly different propositions because their truth-conditions are different. The proposition expressed by (a) is true only if the Nazis' annexing Bohemia is earlier than the time the sentence is uttered, and it is true even if the propositions expressed by (b) and (c) are false. The proposition expressed by (b) is true only if the Nazis' annexing Bohemia is simultaneous with the utterance of (b), and it is true even if the propositions expressed by (a) and (c) are both false. The proposition expressed by (c) is true only if the Nazis' annexing Bohemia is later than the utterance of (c), and it is true even if the propositions expressed by (a) and (b) are both false. The truth of each of these propositions is logically independent of the truth of the remaining two. But given Quine's proposal, however, (a), (b), and (c) must each be equivalent to the same tenseless sentence, *viz.*,

(d) The Nazis annex [tenseless] Bohemia in 1939.

Since (a), (b), and (c) express different propositions they cannot be equivalent to the same sentence.

There cannot be any genuinely tenseless indicative sentences, since there cannot be an indicative sentence without a verb and 'tenseless verb' is a contradiction in terms. So the verb in every indicative sentence must have *some* tense. If a verb must have some tense, then the proposition expressed by an indicative sentence is true only if the proper relation holds between what the proposition asserts and the time when it is asserted. Since the proposition may be asserted at various times, the proper relation need not always hold. So Quine's proposal cannot be used to make it meaningless to ask when a proposition is true.

(3) Just as it makes no sense to apply spatial predicates to truth and falsity, it makes no sense to apply temporal ones to them either.[14]

Bradley is right in claiming that it makes no sense to apply spatial predicates to truth and falsity. Of course a logically contingent proposition might be *true of* a state of affairs which exists in a particular place. But it still makes no sense to suppose that the same proposition might be *true in* one place and false in another. It is nonsense, for example, to suppose that the proposition that the Nazis annexed Bohemia in 1939 is true in Bohemia but false in Brittany. So it makes no sense to ask where (in what place) a proposition is true.[15]

Similarly, even if a logically contingent proposition were *true of* a state of affairs which exists at a particular time, it is (Bradley invites us to conclude) nonsense to suppose that the same proposition might be true at one time and false at another. But this is wrong. It is not nonsense to suppose that the proposition that the Nazis annexed Bohemia in 1939 is true after but not before 1939. Before 1939 it would have been inappropriate to refer to the Nazis' annexation of Bohemia in 1939 using the past tense. So it does make sense to ask when a proposition is true.

We must be prepared to say more than just that it makes sense to ask when a proposition is true. In addition we must define what it is for the proposition expressed by a (tensed) sentence to be true when the sentence expressing it is uttered.

Giving such a definition is complicated once we have rejected the thesis that there can be tenseless sentences, unfortunately. We cannot use a perfectly general one like this.

> The proposition expressed by a sentence of the form 'E occurs [tenseless] at t_1' is true at $t_2 =_{\text{df}}$. Conditions sufficient for its being the case that E occurs at t_1 obtain at t_2.

Instead we must settle for a 'family' of definitions using only tensed verbs throughout. One such definition is (D1), a definition of what it is for a sentence in the present tense to express a proposition which is true when that sentence is uttered.

(D1) The proposition expressed by a sentence of the form 'E now occurs' is true *when that sentence is uttered* $=_{\text{df}}$. Conditions obtain when that sentence is uttered which are sufficient for E's occurring then.

The proposition expressed by the present-tense sentence, 'The Nazis annex Bohemia' is true when that sentence is uttered, for example, if and only if conditions then obtain which are sufficient for the Nazis' then annexing Bohemia. That proposition would not be true then if any conditions necessary for the Nazis' then annexing Bohemia were not satisfied then. The state of affairs which is logically sufficient for that proposition's being true then are the Nazis' annexing Bohemia at the same time the sentence, 'The Nazis annex Bohemia', is being uttered.

We have already noted that the same sentence with the indexical, 'now', can express as many different propositions as there are times when that sentence can be uttered. 'Now' can refer to any particular time only once, i.e., when that time is the present. So the true proposition that an event occurs at *that* time can be expressed by a present-tense sentence only once. (Later on to say that E *occurred* at that time, which has become a time in the past by then, we must use a different sentence, one in the past tense, to express the different proposition, *viz.*, the proposition that E occurred at such-and-such a time in the past.) Moreover, even if (D1) is satisfied at some given time, t, it does not follow that the same sentence, uttered at some other time, t' – and hence expressing the different proposition that E now (i.e., at t') occurs – must therefore express a proposition which is true at t'.

Even if (D1) is satisfied at t, and hence the proposition that E now occurs is true at t, this does not entail the truth at t' of the proposition that E *will* occur at t. This is because (D1)' s being satisfied at t does not entail that when the sentence, 'E will occur at t', is uttered at t' it is uttered at a time *before* t. Nor does (D1)'s being satisfied at t entail the truth at t' of the proposition that E *occurred* at t, since (D1)'s being satisfied at t does not entail that t' is *later* than t. Thus, (D1) reflects the independence of truth-value of proposition expressed by present-tense sentences from those expressed by past-tense and future-tense sentences which we noted before.

A definition of what it is for a sentence in the simple past tense to express a proposition which is true at the time that sentence is uttered is given in (D2).

(D2) The proposition expressed by a sentence of the form 'E *occurred* at t_p' is true when that sentence is uttered $=_{\text{df}}$ conditions obtain when that sentence is uttered which are sufficent for E to have occurred at t_p.

In (D2) 't_p' is replaceable by the description of a particular time before the utterance of that sentence. Although 'E occurred at t_p' contains no indexicals, the use of the past tense makes it a necessary condition of the truth of the proposition it expresses that t_p is earlier than the time of utterance. That proposition cannot therefore be true at and before t_p.

Here is an example of how (D2) might be satisfied. The proposition expressed by the sentence 'The Nazis annexed Bohemia in 1939' is now (1979) true, i. e., conditions now obtain which are sufficient for the Nazis ro have annexed Bohemia in 1939. The conditions which now obtain are these: the Nazis' having (in fact) annexed Bohemia in 1939, and 1939's being earlier than the time of utterance (1979).

A definition of what it is for a sentence in the simple future tense to express a proposition which is true when that sentence is uttered is given in (D3).

(D3) The proposition expressed by a sentence of the form '*E will* occur at t_f' is true when that sentence is uttered $=_{df}$. Conditions obtain when that sentence is uttered which are sufficient for E to be going to occur at t_f.

In (D3) 't_f' is replaceable by the description of a particular time after the utterance of that sentence. The use of the future tense makes it a necessary condition of the truth of the proposition that it cannot be true unless t_f is a time when the sentence is uttered. That proposition cannot therefore be true at and after t_f.

Here is how (D3) might be satisfied. The proposition expressed by the sentence 'The Nazis will annex Bohemia in 1939' was true before 1939, i. e., conditions then obtained which were then sufficient for the Nazis' then being going to annex Bohemia in 1939. The conditions which obtained before 1939 were these: the Nazis' (in fact) being going to annex Bohemia in 1939, and 1939's being later than the time of utterance (before 1939). Whether (D3) is in fact ever satisfied will be pursued further in Part III.

It is a consequence of (D1), (D2), and (D3) that if the proposition expressed by a sentence of one of the following forms:

(a) E now occurs.
(b) E occurred at t_p.
(c) E will occur at t_f.

is true at any given time, this does not entail the truth of either of the remaining propositions – which is as it should be.

We are now in a position to show how it makes sense to ask when a proposition is true or false, and so why (M) is false, given the variations in a proposition's truth-value. The proposition expressed by a sentence of the form '*E* now occurs' can be true when it is uttered but not before or after that time. The proposition expressed by a sentence of the form '*E* occurred at t_p' can be true when it is uttered after, but not at or before t_p. The proposition expressed by a sentence of the form '*E* will occur at t_f' can be true when it is uttered before t_f, but neither at nor after t_f. It is not meaningless to ask when a proposition is true so (M) must be false. Thus, if (E) is interpreted as (M), then (E) is false, and the argument for fatalism is unsound.

III

Even if (M) is false because it is meaningful to ask when a proposition is true, it might still be the case that (E) is true when it is interpreted as (E′), which accommodates and so is not refuted by the variations in truth-value of propositions we noted in Part II.

(E′) (a) For any true proposition expressed by a sentence of the form '*E* occurs [simple *present*] at *t* [the present]' it is logically impossible that either
  (i) the proposition that *E* will occur at *t* was not true before *t*, *or*
  (ii) the proposition that *E* occurred at *t* will not be true after *t*, *and*
(b) For any true proposition expressed by a sentence of the form '*E* will occur [simple *future*] at *t*' it is logically impossible that either
  (i) the proposition that *E* occurs at *t* is false at *t*, *or*
  (ii) the proposition that *E* occured at *t* will be false after *t*, *and*
(c) For any true proposition expressed by a sentence of the form '*E* occurred [simple past] at *t*' it is logically impossible that either
  (i) the proposition that *E* occurs at *t* is false at *t*, *or*
  (ii) the proposition that *E* will occur at *t* was false before *t*.

It has often been remarked that we regard the past as being unalterable either now or in the future. So we are inclined to accept (a) (ii), (b) (ii), and (c) (i).

We can account in the following way for our readiness to accept (a) (ii). Suppose that *E* does in fact occur at *t*. There are two conditions which are jointly sufficient for its unalterably being true after *t* that *E* occurred at *t*. The first condition is that *E* did in fact occur at *t*. This makes the proposition that *E* occurred at *t* true. The second condition is that it must be impossible for anything after *t* to change the truth-value of that proposition. Given that time 'moves' in only one 'direction', it is impossible for it to be true after *t* either that *E* (*that* particular event) is now occurring, or that *E* will occur at *t*. Once *E* has occurred at *t* and *t* has become a time in the past nothing can change the fact of *E*'s occurring or the unique appropriateness of the past tense in a sentence expressing the proposition that *E* occurred at *t*, no matter when after *t* that sentence is uttered. Similar reasons, *mutatis mutandis*, justify our accepting (b) (ii) and (c) (i).

If we are to reject (E') we must be prepared to refute the fatalist's contention that for similar reasons we ought to accept (c) (i) as well. Suppose, again, that *E* does occur at *t*. There are, the fatalist argues, two jointly logically sufficient conditions for its being unalterably true before *t* that *E* will occur at *t*. First, *E* will in fact occur at *t*. This is what makes the proposition that *E* will occur at *t* true. Secondly, it must be impossible for anything before *t* to have changed the truth-value of that proposition. The sentence '*E* will occur at *t*' must always – when it was uttered before *t* – have had the appropriate tense. Given that it is true before *t* that *E* will occur at *t*, then that proposition must always have been true before *t*. Similar reasons, *mutatis mutandis*, justify our accepting (b) (ii) and (c) (i).

We can reply to the fatalist in at least two ways. First, we can point out that our inclination to accept (a) (ii) is nothing more than an inclination to assume, without proof, that whatever is now occurring will in the future be unavoidable. In addition, we accept (b) (ii) only because we are willing to assume, without proof, that whatever will occur will thereafter be unavoidable. Similarly, we accept (c) (i) only because we are willing to assume, without proof, that whatever occurred in the past was thereafter unavoidable. Each of these is a conjunct of the thesis of fatalism, which is the conclusion of the argument in which (E′) is used as a premise. What is happening here is that we are allowing the fatalist to beg the question against us. We may begin to worry finally when we notice that the remaining three are not at all evident to the nonfatalist. In order to accept (a) (i), for example, we must assume, without proof, that whatever is now occurring was always unavoidable, i.e., that no one and no thing could ever have made it the case that *E* does not occur at *t*. In order to accept (b) (i) we must assume, without proof, that whatever has occurred was always unavoidable, i.e., that no one and no thing could ever have made it the case that *E* did not occur at *t*. In order to accept (c) (ii) we must assume, without proof, that whatever will occur was always unavoidable, i.e., that no one and no thing could ever have made it the case that *E* will not occur at *t*. The problem is that the truth of fatalism is necessary for the truth of (E′). So the argument begs the question when (E) is interpreted as (E′). Such an argument can persuade the unwary nonfatalist or the latent fatalist to accept fatalism.

If (E′) does beg the question in this way, then we can refute fatalism by refuting (E′). Before we attempt to do so, however, we shall look at two objections to (E′) which look promising but do not prove it false.

The first objection can be constructed from an argument given by C. D. Broad in *Scientific Thought*. He argues that

> The fact to which a judgment refers is the fact that renders it true or false . . . judgments which profess to be about the future do not refer to any fact, whether positive or negative, at the time when they are made. They are therefore at that time neither true nor false. They will become true or false when there is a fact for them to refer to; and after this they will remain true or false, as the case may be, for ever and ever.[16]

The reason that a proposition which is ostensibly about a future event fails to "refer" to any state of affairs which makes that proposition true or false is that ". . . the future is simply nothing at all".[17]

It might seem that if no sentence of the form '*E* will occur at *t*' can express a true or a false proposition (or if a sentence of that form cannot express a proposition at all), then condition (a) (i) must be false. Even if the proposition that *E* occurs at *t* is true, it is logically possible that the proposition that *E* will occur was not true before *t*. Either there is no such proposition, or if there is one then it cannot possibly be true. For similar reasons, neither (b) (i) nor (c) (ii) can be true either.

This cannot be used to refute (E′), however. Even if the fatalist were willing to accept Broad's claims that (a) what makes a proposition true or false is some fact (or state of affairs), and (b) no future fact 'exists' (or has yet occurred), the fatalist would still argue that there is a state of affairs which does obtain now which makes the proposition that *E* will occur at *t* true (now), *viz*., that *E will* occur at *t*. So it can be true now that *E* will occur at *t*, even if that event has not yet occurred. (If *E* had already occurred, then it would be false that *E will* occur at *t*.)

We can construct an even more serious objection to (E′) – and hence to fatalism – by using an argument given by Gilbert Ryle. He argues not merely that the proposition that *E* will occur cannot be either true or false, but that '*E*' cannot even refer to a particular future event. What he says is this.

. . . statements in the future tense cannot convey singular, but only general propositions, where statements in the present and past tense can convey both. More strictly, a statement to the effect that something will exist or happen is, in *so far*, a general statement. When I predict the next eclipse of the moon, I have indeed got the moon to make statements about, but I have not got her next eclipse to make statements about.[18]

Thus, the proposition that E will occur at t cannot be true unless E is a type of event rather than a particular event. We "have not got" that particular event, so there is no *it* such that the proposition that *it* will occur at t is now true. So condition (a) (i) must be false. Even if the proposition that E (a particular event) occurs at t is true, it is not logically impossible that the proposition that E will occur was not true before t, since there can be no such proposition when 'E' must refer to a particular event. Conditions (b) (i) and (c) (ii) must also be false for similar reasons, and so (E′) must also be false.

What Ryle says cannot be used to refute (E′) without begging the question against the fatalist, whose point is that in the relevant sense we *have* "got" a particular future event before it occurs. To say of a particular event that it is a future event is not to say that it does not exist or that it is 'indeterminate', but only that its time to occur has not yet come. We cannot 'point' to it, of course, but then we cannot 'point' to a past event whose time to occur has already passed. There is no more reason to deny that the proposition expressed by a sentence of the form 'E will occur at t' refers to a particular event than there is to deny that the proposition expressed by a sentence of the form 'E occurred at t' refers to a particular event. To assert that we have not "got" E before it occurs at t is to deny – but not to prove the falsity of – the fatalist thesis that particular future events which will occur are ones which will occur unavoidably.

If we are to reject (E′) it must be on other grounds. To see what those grounds might be we must look further into the question, 'What makes a proposition which asserts that a particular future event will occur at a given future time (t_2) true at some particular time (t_1) before that event is supposed to occur?' More briefly

(QF) What makes it true at t_1 that E will occur at t_2?

The first thing to notice is that we cannot answer (QF) the way we can answer the analogous question about a past event,

(QP) What makes it true at t_3 that E occurred at t_2?

Here is why. The answer to (QP), we might expect, is something like this.

(1P) It is true at t_3 that E occurred at $t_2 \equiv E$ did (in fact) occur at t_2.

E did in fact occur at t_2 only if t_3 is later than t_2. Two things are clear. First, that E *did in fact occur* at t_2 is necessary for its being true later that E occurred at t_2. Secondly, we are not at all tempted to suppose that conditions which exist at t_3 are *causally* necessary or sufficient, or both, for the occurrence of E at the earlier time, t_2. What settles the matter of the truth at t_3 of the proposition that E occurred at t_2 is what happened at t_2 (the occurrence of E) and the temporal relation between t_2 and t_3 (i.e., that t_2 was earlier than t_3). Whatever conditions obtain at t_3 have nothing at all to do with whether that proposition is true.

Contrast this with the parallel answer to (QF), however.

(1F) It is true at t_1 that E will occur at $t_2 \equiv E$ will (in fact) occur at t_2.

It can be shown, however, that E's occurring at t_2 is neither sufficient nor necessary for its being true at t_1 that E will occur at t_2.

E's occurring at t_2 is not sufficient for its being true at t_1 that E will occur at t_2 because E's occurring at t_2 is not sufficient for its being the case that t_1 is earlier than t_2, and hence that what is true at t_1 is the proposition that E *will* occur at t_2.

What is more surprising, perhaps, is that E's occurring at t_2 is not even a necessary condition of its being true at t_1 that E will occur at t_2. The reason that

(2F) E's occurring at t_2 is necessary for its being true at t_1 that E will occur at t_2

is false is that E's occurring at t_2 comes too late to satisfy a necessary condition of its being true before t_2 that E will occur at t_2. Indeed E's occurring at t_2 is sufficient for its being *false* that E *will* occur at t_2, since that event is at t_2 no longer a future event. So either (2F) is false, or else it cannot be true at t_1 that E will occur at t_2.

Perhaps it might be thought, however, that (2F) is true and that that necessary condition can be satisfied retroactively (i.e., met at t_1 by later being met at t_2). There are two objections to this. First, a fatalist cannot accept this consistently with his claim that the past is (subsequently) unalterable. (In effect an event's occurring at t_2 would make something have been the case at t_1.) Secondly, this proposal is self-contradictory. If that necessary condition could be satisfied retroactively, then it would be the case that at t_1 the condition is both unmet (because E has not yet occurred) and met (because E later occurs at t_2). So that necessary condition cannot be met retroactively. It must therefore be the case that either (2F) is false, or else it cannot be true at t_1 that E will occur at t_2.

Thus either it cannot be true at t_1 that E will occur at t_2, or the occurrence of E at t_2 is neither sufficient nor necessary for its being true at t_1 that E will occur at t_2. So we cannot answer (QF) by saying that the occurrence of E at t_2 is part of what makes it true at t_1 that E will occur at t_2.

Again in contrast with our answer to (QP) we seem to be forced to hold that if it is true at t_1 that E will occur at t_2 it is not because of conditions which exist at t_2 (the occurrence of E) but rather because of conditions which exist at t_1, as in

(3F) Conditions existing at t_1 are sufficient for its being true at t_1 that E will occur at t_2.

We can describe those conditions trivially as being 'conditions which exist at t_1 and which are sufficient at t_1 for its being the case at t_1 that E is going to occur at t_2'.

No conditions which exist at t_1 can be either logically sufficient or causally sufficient for its being true at t_1 that E will occur at t_2. Let us see why.

No conditions which exist at t_1 can be *logically* sufficient for its being true then that E will occur at t_2 because there can be no proposition about conditions existing at t_1 which *entails* that E will occur at t_2. The truth of some proposition about conditions existing at t_1 is logically consistent with E's not occurring at t_2. This is because even if those conditions do obtain at t_1 there is no contradiction in supposing that a miracle or some other unforeseen state of affairs should occur after t_1 which prevents E from occurring at t_2.

No conditions which exist at t_1 can be causally sufficient for its being true at t_1 that E will occur at t_2 either. *The* conditions which are causally sufficient for the occurrence of E at t_2 consist of all of the conditions which occurred before t_2 or which occur at t_2 and which together bring about the occurrence of E at t_2. Since some of these conditions must exist after t_1 but before t_2, some of the components of *the* sufficient causal condition for E's occurring at t_2 do not yet exist at t_1. Moreover, some of these components existed before t_1 but have ceased to exist by then. These are part of the sufficient causal condition for E's occurring at t_2, yet they no longer exist at t_1. Thus, the conditions existing at t_1 cannot be *the* sufficient causal condition for E's occurring at t_2.

We must conclude, then, that at t_1 it cannot be true that E will occur at t_2. It is impossible for there to be any conditions which exist either at t_1 or at t_2 which make it true at t_1 that E will occur at t_2.

What follows from this is that (E′) is false. Its conjunct (a) is false because even if a sentence of the form '*E* occurs at *t*' is true the proposition that *E* will occur at *t* was still not true before *t*. So it is false that for any true proposition expressed by a sentence of the form '*E* occurs at *t*' it is logically impossible that the proposition that *E* will occur at *t* was not true before *t*. For similar reasons its conjuncts (b) and (c) are also false, because of the falsity of (b) (i) and (c) (ii) respectively. If (E′) is false, and if fatalism is true only if (E′) is true, then fatalism is false.

Syracuse, N.Y.

NOTES

[1] See especially Richard Taylor, 'Fatalism', *Philosophical Review* **71** (1962), 56–66; 'Fatalism and Ability', *Analysis* **23** (1962), 25–27; "Diodorus Cronus" (i.e., Richard Taylor and Steven M. Cahn), 'Time, Truth and Ability', *Analysis* **25** (1965), 137–141.

[2] Richard Taylor, *Metaphysics*, 2nd ed., Prentice-Hall, Englewood Cliffs, N.J., 1974, pp. 58–71.

[3] *Ibid.*, p. 59.

[4] *Ibid.*

[5] *Ibid.*, p. 65.

[6] I owe this point to Professor José Benardete.

[7] R. D. Bradley, 'Must the Future Be What It Is Going to Be?', *Mind* **68** (1959), 193–208. Reprinted in Richard M. Gale (ed.), *The Philosophy of Time: A Collection of Essays*, Anchor-Doubleday, Garden City, N.Y., 1967, pp. 232–251. The page numbers here refer to the Gale volume.

[8] *Ibid.*, p. 241. A. J. Ayer defends this thesis in 'Fatalism', in *The Concept of a Person and Other Essays*, St. Martin's, New York, 1963, pp. 235–238.

[9] Bradley, *op. cit.*, p. 241.

[10] *Ibid.*, p. 242.

[11] Willard Van Orman Quine, *Elementary Logic*, rev. ed., Harvard University Press, Cambridge, Mass., 1966, p. 6. The locution "true once and for all" might be misleading, since it apparently allows the question, 'When is *p* true?' to be meaningful, since it can be answered, '*p* is true *once and for all*'. Following a suggestion by Peter van Inwagen, perhaps it would be clearer to use 'true (if true) and false (if false) – independently of time'.

[12] Bradley, *op. cit.*, p. 243.

[13] Richard Taylor says in 'The Problem of Future Contingencies', *Philosophical Review* **66** (1957), 18, that "... in fact one *cannot* convey the same information and avoid the systematic ambiguity of 'now' and 'then' just by substituting dates for tenses. This can be done *only* if an additional statement is supplied in order to complete the information so easily completed by the use of tense, and this additional statement must contain a temporal reference to *now*. That is, it must be a statement to the effect that the date

mentioned is earlier than, contemporaneous with, or future to *now* – precisely the thing that the use of dates was intended to avoid". This seems absolutely correct.

See also Steven M. Cahn, *Fate, Logic, and Time*, Yale University Press, New Haven, Conn., 1967, pp. 34–35, 127. He agrees with Taylor that it is impossible to translate a tensed sentence into a tenseless one without loss of meaning, but he shows this on epistemological grounds. One can know the 'fact given' in a tenseless proposition without thereby knowing the fact given in the original tensed one, so the fact given in the original tensed one has not been preserved by the translation.

14 Bradley, *op. cit.*, p. 244.

15 It is suggestive that there is no temptation to translate 'placed' sentences into 'placeless' ones to insure that it makes no sense to ask where the propositions expressed by the placeless ones are true.

16 C. D. Broad, *Scientific Thought*, Routledge and Kegan Paul, London, 1923, pp. 70, 73.

17 *Ibid.*, p. 66.

18 Gilbert Ryle, *Dilemmas*, Cambridge University Press, Cambridge, 1954, p. 27.

D. M. ARMSTRONG

IDENTITY THROUGH TIME*

Some philosophers in their work are led on to ever greater complexity; others seek simplicity and clarity of argument and vision. Each type of mind serves to check the shortcomings of the other. In our age philosophy is more professionalized than ever before, so as a result the first sort of mind is in the ascendant. All the more important, therefore, is the role of those who will not let their thought be dissipated in endless ramifications. Richard Taylor's particular intellectual contribution has been to discover, or to restate, simple and direct, yet profound and forceful, arguments which lead to important conclusions about major philosophical issues. He has done this in a way which involves no sacrifice of contemporary standards of rigor and exactness.

I. IDENTITY THROUGH TIME: THE IDENTITY AND THE RELATIONAL VIEW

Two views may be taken of the identity of particulars through time. We may call them the identity view and the relational view. Many, but not all, contemporary analytical philosophers accept the relational view. So do I. Recently, however, as a result of working on the problem of universals, I have come to have more sympathy with the identity view. I still think it is false, but I do not think, as I used to think, that it is nonsense. In this paper I want to consider again the dispute between the two views.

It is to be understood that neither of these views challenges the truism that if X and Y are different phases of an object, P, then they are phases of one identical thing. The identity and the relational views are, rather, different philosophical analyses of the situations referred to in the previous sentence. I oppose an "identity" to a "relational" view because I believe that identity is not a genuine relation. If this is an incorrect belief, however, for our present purposes it means no more than that my terminology is ill-chosen.

I begin by trying to characterize the relational view. If we consider non-overlapping phases in the history of the same particular, P, then, according to the relational view, such phases are in no way identical. What we have is simply a particular case of different *parts* of the same thing. They are temporal parts rather than spatial parts, but the adjectives 'temporal' and 'spatial'

Peter van Inwagen (ed.), Time and Cause, 67–78.
Copyright © 1980 *by D. Reidel Publishing Company.*

do not modify the meaning of the word 'parts'. These parts, themselves particulars, are related in various ways to each other and to further particulars. The holding of some of these relations *constitutes* what it is for the parts to be different temporal parts of *P*.

Consider, by way of comparison, nonoverlapping spatial parts of *P* at a certain instant of time. Very few philosophers would want to argue that such spatial parts are in any way identical with each other. These parts, themselves particulars, are related in various ways, to each other and to further particulars. The holding of some of these relations *constitutes* what it is for the parts to be different spatial parts of *P*. Just what these constituting relations are will depend upon what sort of thing *P* is. The relations between the handle of a cup and the rest of the cup which make the two particulars into a cup are rather different from the relations which make the soldiers of an army into an army. But the unity of *P* at a time seems secured simply by uniting relations between *P*'s spatial parts at that time, or, in some cases, their relations to some further particular or particulars.

If we accept this view about the spatial case, it is natural to conclude that the temporal parts, that is, the phases, of *P* are also united to form *P* by nothing more than relations.

Nevertheless, we do have a sense that nonoverlapping phases of the history of the one particular are not mere different parts of the same particular, but are actually *identical* with each other in some deeper way. It is the same thing which existed yesterday and today, we are inclined to think, in a way that the different spatial parts of the same thing are not identical with each other in any way. We do have some intuitive sense that a relational analysis of identity through time is false and that an identity theory is true.

Is this feeling just nonsense, nonsense which, perhaps, appeals to us because of our deep emotional interest in the continuing of ourselves and other things which we cherish? I think that Hume thought it was nonsense, although he recognized that it was nonsense to which we are instinctively attracted. He spoke of the way in which we 'feign' an identity between the different phases of the same thing.[1]

A strong argument can be advanced in support of Hume's view that the identity account of identity through time is nonsense. Consider again two nonoverlapping phases of the particular *P*. However intimately they may be related, it seems impossible to deny that they are two wholly distinct particulars. This becomes clear when we consider that we can attribute properties and relations to both phases, and that these properties and relations may be different from (and even incompatible with) each other. *P* may be at one temperature during the first phase, another temperature during the second. Things which differ in their properties are different things. But if the two phrases are different things, then they are not the same thing.

I think that this argument does show that nonoverlapping phases of the same particular are distinct particulars. What it fails to show, however, is that they are *wholly* distinct particulars. It leaves open the possibility that they are distinct, but not wholly distinct.

In thinking about identity there is a tendency for one's thought to be dominated by the two extreme cases. These cases may be illustrated by the complete identity of the morning star with the evening star, on the one hand, and the complete nonidentity of the morning star with the Red Planet, on the other. But we need to remember that there are intermediate cases: for instance, terraced houses which have a party wall in common, or unseparated Siamese twins. Another sort of case is that where phases of the same particular overlap but do not coincide completely. Wherever things overlap or stand as part to whole we have merely partial identity.

But what is the relevance of all this to the case, which we are considering, of *nonoverlapping* phases of a particular? The answer is that, given certain further assumptions, even different particulars of this sort may be partially identical. Suppose, in particular, that one rejects nominalism, the doctrine that nothing exists which is not a particular. Suppose that, as I do, one accepts the objective reality of (instantiated) universals. Suppose one holds that different particulars may have the very same, the *identical*, property, for instance, the same mass. One is then committed to saying that such particulars are (at least partially) identical in nature. And if identity is a univocal notion, as I believe it is, partial identity of nature entails partial identity.

To think along these lines is to have one's ideas about the complete nonidentity of nonoverlapping particulars shaken up in a fruitful way. But it is hardly to accept that nonoverlapping phases of the same particular could be identical *as particulars*.

To understand this latter possibility we need to consider what a particular is. The view which I accept is the orthodox view among those who are realists, that is, are believers in universals. A particular is, essentially, a particular-having-certain-properties-and-relations. Its particularity is distinguishable but not separable from its properties and relations. Following Scotus, we can say that it involves both thisness and nature, or with Aristotle that it is a this-such.

Against the background of this view we can now characterize the identity analysis of the identity of particulars through time. We may note as a preliminary that identity analyses are normally restricted to certain favored particulars, such as atoms and spiritual substances.[2] Suppose that *X* and *Y* are nonoverlapping phases of a particular *P*. *X* and *Y* are particulars themselves, particulars having certain properties and relations. The identity view is that the particularity or thisness of *X* is identical with the particularity or thisness of *Y*. The relational view is that they are distinct.

My present inclination is to say that both identity and relational analyses are intelligible hypotheses. I reject the identity analysis, looking rather to relations between different phases to secure the unity of a particular over time. But I do not think that the identity view can be rejected as illogical. If it is to be rejected, then I think it must be rejected for Occamist reasons. The different phases exist, and so do their relations. These phases so related, it seems, are sufficient to secure identity through time for all particulars.

II. LOCKE ON IDENTITY THROUGH TIME

I suggest, then, that the identity view of identity through time is not illogical. The question is rather whether it is a postulation which is fruitful, or expedient, or which we are compelled, to make. I shall try now to illuminate the difficulties faced by an identity theory by a discussion of Locke's views.[3] Locke, I believe, accepted the identity view, at any rate for certain sorts of entity. I think that this is what he meant when he spoke of "identity of substance": ". . . whatever substance begins to exist, it must, during its existence, necessarily be the same".[4] Such identity appears to be one of the notions involved in the Lockean notion of a *substratum* which underlies and supports the properties and relations of things. Among other things, the substratum is that which is identical in the different phases of a thing's existence. However, this function does not seem to be an essential part of the doctrine of substratum. It is possible to have a less rich conception of substratum which could be combined with a relational theory of identity through time. Contrariwise, it would be possible to take an identity view and yet reject the notion of an underlying substratum. The particularity of a particular would, then, not be identified with a substratum. Locke's substratum, however, both underlies properties and relations and is also identical through different phases. It is the latter function only which we are interested in here.

But there is a difficulty for the identity view, a difficulty of which Locke is well aware:

. . . if two or more atoms be joined together into the same mass, every one of those atoms will be the same, . . . and whilst they exist united together, the mass, consisting of the same atoms, must be the same mass, or the same body, let the parts be never so differently jumbled. *But if one of these atoms be taken away, or one new one added, it is no longer the same mass or the same body.*[5]

From the perspective of a relational view of identity through time this can only seem to be extreme pedantry. For the relational theorist is not saddled with any very strict rules as to what will constitute 'the very same thing at a later time'. The rules may differ for different sorts of things. But Locke has less room for maneuver. On the basis of the science of his day Locke assumes, as we still assume, that ordinary physical objects are made up of atoms. Further he adopts an identity view of their identity through time. (Unlike our atoms, they are physically indivisible.) Now suppose that an ordinary physical object loses a few atoms and gains others over a period of time. The atoms which it loses continue their career elsewhere. What object at the latter time is identical with the object at the earlier time? For an identity theorist the best candidate *must be* the collection of the original atoms with which the object started out, in however scattered a state those atoms are at the later time. There the identity is complete. By comparison, the collection of atoms actually adhering together is only *partially* identical with the original collection. This, I believe, is why Locke wrote as he did in Section 4. The difficulty he is in is a difficulty only for one who holds an identity view. This in turn is evidence that, with respect to atoms, Locke did hold an identity view.

The situation is still more paradoxical for Locke in cases of objects all of whose matter is replaced over a period of time, as is said to occur in the case of the human body. Since the body in its later state does not contain a single one of the original atoms, it cannot even be partially identical with the body in its earlier state. They are two completely different substances.

What is Locke to do? He is well aware of this problem faced by an identity theory. He wants to give a nonskeptical account of identity through time for human and animal bodies, trees and so on. One solution for him would be to fall back on a relational analysis for such objects. And, indeed, when it comes to giving an account of *personal* identity Locke does embrace a relational analysis. In the case of the bodies of animals and men, however, his view appears to be different. He says of the bodies of animals that: ". . . animal identity is preserved in identity of *life* and not of substance."[6] What is meant by "identity of life"? Locke, I believe, thinks of life here as a certain property which the animal possesses. In particular, it is a structural property, a way in which the (spatial) parts of the animal are related and organized.[7] Locke calls such a property a "mode", and, when summing up his position, after having said that: ". . ; whatever substance begins to exist, it must, during its existence, necessarily be the same; . . ." he also says that: ". . . whatsoever mode begins to exist, during its existence it is the same".[8]

I think that what he is suggesting here is an identity theory for properties as well as substances. Suppose that particular *a* has property *L* at t_1 and that *b* has *L* at t_2. Suppose, also, that *a* and *b* are not even partially 'identical in substance'. It may nevertheless be the case that *L* at t_1 is identical with *L* at t_2. If so, then it can be said that *a* and *b*, although not different phases of the same *substance*, are different phases in the existence of the same *L*. Suppose *L* to be the property of *being a living thing*. Then *a* at t_1 and *b* at t_2 are different phases in the existence of the same living thing. It might have been a caterpillar at t_1 and a butterfly at t_2.

It seems that a restriction will have to be put upon the sort of property which *L* is. It will have to be the sort of property which 'divides its instantiations' so that, at a given time, there is a certain definite number, finite or infinite, of instances of *L* in the universe. But, given this restriction to the sort of property *L* is, may not Locke postulate such an identity to account for the identity through time of such things as oaks?

There is, however, a difficulty which Locke must face here. If *L* is a property, and a property is a universal, then it is all too easy for *L* to be identical with itself at different times. For it is the same property wherever it is instantiated. This has the consequence, contrary to what Locke requires, that no special identity is set up between *L*, the property of *a* at t_1, and *L*, the property of *b* at t_2.

What this makes clear is that the conception of a property which Locke requires if he is to make good his notion of property identity through time is not that of a property as a universal but that of a property-instance. Certain philosophers, G. F. Stout was a prominent Anglo-Saxon example, hold that where two billiard balls are red or are spherical each of the balls has its own redness or sphericity. These properties are not universals, but are as particular as the billiard balls themselves. This position may or may not be combined with the admission that these particular properties themselves have universal properties ('This redness is red').

Let it now be given that the *L*-ness of *a* at t_1 and the *L*-ness of *b* at t_2 are particulars. Might they not be different phases of the same property-instance? And might we not give an identity rather than a relational account of what makes the two phases phases of the same property-instance? It is difficult to see how the distinction between particularity and nature could be drawn in the case of property-instances. But perhaps it could be said that the different phases in the existence of a property-instance are not numerically diverse, as a relational theorist would assert, but are instead completely identical.

When Locke asserts that: "whatsoever mode begins to exist, during its existence it is the same", he appears to be asserting that we must give an identity analysis of the situation. I do not see that an identity analysis is required, but granted the conception of a property-instance, it seems to be a possible analysis.

Whether or not Locke explicitly holds to the notion of property-instance, I do not know. But it may be noted that it would fit in well with his general position. Locke talks freely of modes and qualities. Yet at the same time he is an explicit nominalist, holding that everything there is is a particular. If he is to be consistent, then he should hold that modes and qualities are particulars. And we have seen that this doctrine is required if he is to have any show of maintaining his doctrine of the identity of modes over time.

Locke's building, however, is only as strong as its foundations. And here I am given pause. It would take me too far afield to argue the matter in detail, but I am convinced that the notion of a property-instance turns out in the end to be an incoherent one. The idea of a property is the idea of a nature, a nature which is different for different properties. I do not think that we can make any sense of the notion of nature unless we conceive of properties as universals. If this is correct, then Locke's ingenious attempt to supplement his identity view of the identity of substances through time with an identity view of the identity of modes through time, does not succeed.

III. THE RELATIONAL VIEW DEFENDED

Relational theorists are inclined to argue that identity theories of identity through time are unintelligible. I have argued, however, that identity views are in good logical order. Perhaps postulating such identities is to fall into an abyss of nonsense but I cannot at present see that this is so. I have argued that Locke's notion of the identity of modes through time rests upon the notion of a property-instance, and the latter notion, I believe, is an untenable one. But if property-instances are acceptable, as I think they are not, then perhaps an identity theorist can even deal with the identity through time of things which 'change their substance' during that time.

But to show that the identity theory is in good logical order is not to show that we should adopt it. We can distinguish between, and refer to, different phases in the existence of the same particular. Since the different phases will, in general at least, have different properties, we must recognize them as different particulars. Such difference is compatible with partial identity, and to the extent that particulars have common properties they are partially identical. But why should we say that different phases of a particular are identical in their *particularity*? It is clear from our discussion of Locke's difficulties that, at best, it is plausible to say this only of certain favored individuals which do not change their substance, such as Newtonian atoms. But even in such a case it seems that unity can be secured for the enduring individual simply by relations between the phases. To extend the relational analysis to all enduring particulars is therefore economy of theory.

As already mentioned in Section I, one very powerful intellectual motive here is our unwillingness to contemplate an identity view with respect to the different spatial parts of the same thing. As Hume points out in his satirical way, in spatial cases we find the scholastic maxim, "*totum in toto, et totum in qualibet parte*" [the whole in the whole and in each part], quite unacceptable.[9] I should still want to argue, against Hume, that this identity view is, or can be presented as, an intelligible hypothesis. But surely an adequate account of spatial unities can be given without appealing to anything except the relations which the spatial parts of the object, be it nation, plant, animal, or stone, bear to each other? Any other account seems to be extravagance. But if a relatonal account is possible in the case of spatial unities, that is surely a strong reason for thinking that a similar account can be given of the different *phases* of the same thing. To deny the parallel is to set up a very dubious distinction between space and time.

I note here in passing that it has been a major theme in Richard Taylor's metaphysical reflection to reject various asymmetries between space and time. In this particular case, to combine an identity view of identity through time with a relational view of identity across space seems even to involve scientific difficulty. For how is the combination to be effected if the simultaneity of spatially separated things is a relative rather than an absolute notion (a three-termed rather than a two-termed relation)?

If we are looking for relations to bind together the spatial parts of a thing so that it constitutes a thing of a certain sort, then, in general, we must appeal to different relations in the case of different sorts of things. But there is one relation which seems to be of quite peculiar importance in the case of the spatial parts. It is causation. In the case of solid objects, and particularly in the case of organisms, reciprocal causal relations between the spatial parts are all-important. By contrast, in the case of identity through time it is fashionable to set great store by spatiotemporal continuity. Again, however, I believe that we ought to set greater store by causal relations.

Hume was well aware of the importance of causality for relational accounts of identity through time. Ironically, it is probably the ontological downgrading of causality involved in Humean analyses of causality which has brought about the modern neglect of causal relation as a cement for different phases of the same thing. In the case of spatial parts the causal relation stares us in the face, so that it cannot be ignored.

I reject all Humean analyses of the causal relation. I do accept the Humean view that causal connection is a species of nomic connection. But I reject the view that nomic connection is nothing more than a regularity. In my view it is a regularity determined by a relation between universals. Causal connection *so conceived* seems capable of welding different phases of the same particular much more closely together than mere spatiotemporal continuity.

We do not normally speak of an earlier phase of an object as being the cause of a succeeding phase. But, in general at least, the earlier phase will be one of the nomically necessary conditions for the existence of the latter phase. Quite often, it will be absolutely nomically necessary. For instance, there is no known method, at least, of creating an adult human being except by creating an embryo and letting it develop under appropriate conditions. Embryonic and childhood phases appear, at least, to be absolutely nomically necessary for the existence of adult phases. But even where there is no such absolute nomical necessity, the earlier phases of an object may still be nomically necessary for later phases in a given particular situation. Given the concrete situation, the recent existence of this desk I write on is nomically necessary for the current existence of this desk. For consider the concrete situation which obtained in this room a few minutes ago, but subtract from it the desk. It is nomically impossible that in that situation a desk should come to be in my room now having the same properties as the original desk. In all probability, it is nomically impossible that in that situation a desk should come to be in my room now having rather similar properties to my desk.

So we seem justified in saying that, for the vast majority of cases at least, preceding phases of a thing are a necessary part of the total cause which brings the succeeding phases to be. The succeeding phases are got by way of the preceding phases, even if for many things (plants and animals especially) much cooperation from the environment is also needed. All this paves the way for the suggestion that, for most sorts of things at least, this causal relation between phases is a logically necessary condition for the *identity* of that thing through time.

The suggestion may be supported by considering a case where the usually suggested marks of identity through time, such as spatiotemporal continuity, are present but it is given that there is no causal connection between an earlier and a later 'phase'. (I call this method of arguing for a logically necessary condition 'the method of subtraction'.) Suppose, then, that there are two very powerful deities, each able to annihilate and create, who operate quite independently of each other. The first deity decides to annihilate Richard Taylor and does so that place *p*, time *t*. The second deity has not been watching what was happening. He decides to create a man at *p* and *t*. By a coincidence which can only be described as cosmic, he decides to give this man exactly the same physical and mental characteristics that Taylor had at *p* and *t*. Life goes on as usual.

The question is 'Did Taylor survive?', 'Is Taylor_2 identical with Taylor_1?' I hope that the reader will agree with my intuition that he did not and is not. It is true, of course, that everybody earthly, including Taylor_2, will take for granted (because they will not even raise the question) that Taylor_2 is Taylor_1. But I think that that only shows that, given the right stage setting, it is logically possible that we should be deceived about anything at all, including Taylor about Taylor's identity.

We may note, incidentally, that a spatiotemporal gap seems quite unimportant provided the proper causal connection is present. If Taylor_1 appears to be annihilated at t_1 and p_1, and Taylor_2 comes into existence at t_2 and p_2 as, or much as, Taylor_1 was at t_1 and p_1, and if further the coming-to-be of Taylor_2 stands in a suitable causal relation to Taylor_1, then Taylor_2 would appear to be simply a later phase of the existence of Taylor_1. (It will be noticed that I say 'suitable' causal relation. I am far from thinking that I have adequately characterized the particular nature of the causal relation which holds between different phases of the same thing.) Spatiotemporal continuity of phases of things appears to be a mere result of, an observable sign of, the existence of a certain sort of causal relation between the phases.

I will finish this paper by considering another imaginary case which I think brings out quite strongly the attraction of an identity view of identity through time. But I will then argue that the puzzle which it creates for the relational view can be resolved by an appeal to causation.

In general, we can distinguish between a spherical object rotating on its own axis and that same object stationary. Consider, however, an exactly spherical object of completely homogeneous material containing no empty space at all. (Perhaps the situation is no more than logically possible.) Can a relational theory make the distinction between such a sphere stationary and the same sphere rotating on its own axis? Let it be granted that there is no way of telling whether rotation is occuring or not. It seems natural still to say that it might rotate or not rotate.

However, the difficulty for a relational theory of identity through time is that if the sphere is considered at any instant, its nature, including its relations to other things, appears to be exactly the same whether the sphere is rotating or not. But if the nature of the sphere supposedly stationary and the nature of the sphere supposedly rotating are the same at every instant, then it seems that, contrary to intuition, no distinction can be drawn between the sphere stationary and the sphere rotating.

It is here that the identity view becomes attractive, as enabling us to draw the distinction. Suppose that the sphere is not rotating. Consider, then, the eastern portion of the sphere from t_1 to t_2. The phases of the eastern portion will not merely remain identical in nature (as they would if the sphere were rotating), but will at all times be phases of the very same thing, in the sense which an identity analysis would yield. If, however, the sphere starts to rotate, then such an identity of the eastern portion will not be maintained except at the instants when the sphere has made exactly N revolutions, for any whole number N. On the relational analysis, however, it seems that there is no difference between the stationary and the rotating sphere.

But I suggest that a relational view can appeal to causality here as a means of differentiating the two cases, though it will certainly have to be causality conceived of in a non-Humean manner. If the sphere is stationary, then the phases of the eastern portion from t_1 to t_2 will bear to each other that particular causal relationship which is required for phases of the same thing to constitute phases of the same thing. In particular, the existence of the earlier phases will be nomically required for the existence of the later phases in a way which will not be so for different temporal phases of spatially separate portions of the sphere. If the sphere is rotating, the causal relations will at once be different.

The point may be brought out by considering the sphere with a segment removed. If the sphere is stationary, the empty segment stays stationary, because nothing comes from nothing. If the sphere rotates, then the empty segment moves round for exactly the same reason. The hypothesis that the sphere is really stationary but that the empty segment moves round is logically possible. But it demands annihilations and creations at every moment of revolution which we believe to be nomically impossible.

So I suggest that we should continue to favor a relational account of identity through time, appealing in particular to the relation of causality between the successive phases. Causality does seem to furnish a powerful enough cement to bond together different phases of the same thing. Only if this program encounters insuperable difficulties need we consider falling back upon an identity analysis.

University of Sydney

NOTES

* I am greatly indebted to Bruce Langtry, Peter van Inwagen, and Sydney Shoemaker for criticism of earlier drafts of this paper.

1 *A Treatise of Human Nature* 1. 4. 2, 6, ed. by L. A. Selby-Bigge, Oxford University Press, London, 1888, pp. 187–218, 251–263.

2 One contemporary philosoher who appears to accept an identity analysis of identity through time, at any rate with respect to the *self*, is Roderick Chisholm. See Section 1 of his 'Problems of Identity' in *Identity and Individuation*, ed. by Milton K. Munitz, New York University Press, 1971, pp. 3–30.

3 Presented in his rich though difficult Chap. 27 ('Of Identity and Diversity') in Book 2 of *An Essay Concerning Human Understanding*, 2 vols., ed. by A. C. Fraser, Oxford University Press, Oxford, 1894, Vol. 1, pp. 439–470.

4 *Essay* 2. 27. 28; Fraser edition, Vol. 1, p. 469.

5 *Essay* 2. 27. 4; Fraser edition, Vol. 1, pp. 442, emphasis is added.

6 *Essay* 2. 27. 12; Fraser edition, Vol. 1, p. 453, emphasis added.

7 Section 5 may be quoted here (Fraser, Vol. 1, p. 443): "We must therefore consider wherein an oak differs from a mass of matter, and that seems to me to be in this, that the one is only the cohesion of particles of matter any how united, the other such a disposition of them as constitutes the parts of an oak; and such an organization of those parts as is fit to receive and distribute nourishment, so as to continue and frame the wood, bark, and leaves, etc., of an oak, in which consists the vegetable life. That being then one plant which has such an organization of parts in one coherent body, partaking of one common life, it continues to be the same plant as long as it partakes of the same life, though that life be communicated to new particles of matter vitally united to the living plant".

8 *Essay* 2. 27. 28; Fraser, Vol. 1 p. 469.

9 *Treatise* 1. 4. 5; Selby-Bigge edition, p. 238.

PART II

CAUSATION

HÉCTOR-NERI CASTAÑEDA

CAUSES, ENERGY AND CONSTANT CONJUNCTIONS

Table of Contents:

INTRODUCTION

Most contemporary philosophers writing on causality proudly declare their adoption of a Humean framework. But many who ostensibly write from an anti-Humean perspective also adopt Hume's main tenets about the nature of causality. This is the case even with Kant and those who believe that we have a definite concept of causal necessity on a par with the concept of logical necessity. To be sure, these opponents of Hume do disagree with him on very important issues. Yet their disagreements with Hume are not for the most part about the structure or nature of causality. The disagreements are, rather, about the origin of the concept of cause, or about the constitution of the mind that has the concept of causality, or, in the most radical cases, about the metaphysical source of causality, but not about the causal connection itself.

For a deeply non-Humean analysis of causation we have to reach backwards into Leibniz's view of causality. Only a Leibnizian-type investigation of causation can help the philosophical discussion of causality move forward. This is precisely what I propose to defend here. This essay is simply a preliminary Leibnizian-type reflection on causation. In spite of its preliminary character, some interesting results begin to emerge. Among others, we can see

Peter van Inwagen (ed.), Time and Cause, 81–108.
Copyright © 1980 *by D. Reidel Publishing Company.*

how causation connects with constant conjunctions of properties in certain (closed) segments of the universe, and how causation connects with energy and local time. We formulate some fundamental principles presupposed in the measurement of causation. Hopefully, a skillful metaphysical logician, or logical metaphysician, pursuing this line of investigation will bring forth the most exciting and valuable clarification of causality through a rich, rigorous, and comprehensive theory of the causal relations. I also hope that some skillful logical historian of philosophy can set the record straight on the contrasts between Hume's and Leibniz's views on causality (*see 'Note Added in Proof' on page 108*).

For convenience we demarcate the general Humean framework as a foil for our 'Leibnizianesque' reflection.

1. HUME'S MAIN TENETS ABOUT CAUSALITY

Evidently, Hume's views about causality are intimately connected with his concept-empiricism (that all our concepts originate in experience), his judgment-empiricism[1] (that there are no synthetic a priori propositions), his phenomenalism, and his nominalism. But we must set all those doctrines aside. His main tenets about causality are independent of those doctrines. These are crucial, of course, for his arguments for his claims about causality. Here we are interested just in Hume's views on the nature of the causal relation, not in his general philosophy. He conceives the causal relation to be characterized by the following main theses:[2]

H1. A cause *c* of an effect *e* is temporally prior to *e*.[3]

H2. A cause *c* of an effect *e* exists at a place that is the same as, or is contiguous with, the place at which *e* exists.[4]

H3. An item *c* is the cause of an item *e*, only if all those items "*resembling the former are placed in a like relation of priority and contiguity to those* [items] . . . , *that resemble the latter*."[5]

H4. An item *c* is the cause of an item *e*, only if *c* and *e* are "*so united . . . in the imagination, that the idea of the one determines the mind to form the idea of the other, and the impression of the one to form a more lively idea of the other*."[6]

H5. A cause *c* is not, strictly speaking, connected necessarily with its effect *e*, since there is only one kind of necessity, namely, logical.[7]

H6. The necessity which we incorrectly attribute to causality, physical necessity, is just the combination of H3 and H4.[8]

We are not concerned with the origin of our concept of cause. Hume may be right in claiming that " 'tis from [our experience of] ... resemblance [as per H1–H3], that the ideas of necessity, of power, and of efficacy, are deriv'd."[9]

Hume was not always very clear on what the causal items are. In the *Treatise* he speaks of objects, but in the *Abstract of the Treatise* and the *Enquiry Concerning Human Understanding*[10] he speaks of causal connections between changes or events. On this point the Humean tradition has tended for the most part to the view that causally related items are events. We still need a deep theory of events. Perhaps that is why some philosophers within the Humean tradition have taken causality to relate states of affairs (or propositions).

H1 and H2 together establish that causality requires *spatiotemporal contiguity* and does not operate at a distance. This is an important condition which is, e.g., violated by Newton's own formulation of the law of gravitation.

H3 is the *constant conjunction requirement*. It applies literally to the past, but together with H4, it can be construed as requiring that causality involve an exceptionless universal constant conjunction, i.e., a constant conjunction that holds at all times. We shall refer to this universal requirement as the *cognitive content* of our concept of causality.

H4 formulates a requirement on our concept of causality that has less to do with what we think when we think that an event *c* causes an event *e*, than with the mental set we get into (cause ourselves to be in, one feels tempted to say) when we think that *c* causes (or caused) *e*. This mental set is very complex. It includes a readiness to infer the future cases resembling *e* will follow cases resembling *c*, etc. Naturally, when we think that *c* is the cause of *e* we have a vague, diffuse, inarticulated consciousness of our appropriate mental set: we acquire, so to speak, a shadowy impression of our mental set. We express that sensorylike consciousness when we assert that something *c* causes something else *e*. Hume did not elaborate this fully; but one can suspect that the crucial thing, given theses H5 and H6, is that the element of vague self-consciousness underlying our thoughts and claims about causes is not vivid (he would have said) enough to constitute an element of the idea or concept of cause. There is no genuine self-reference, only a sensorylike presentation of what we may call the mental 'causal' setting. Thus, in thinking or saying that *c* is the cause of *e* we are, in the present regard, creating or performing, rather than contemplating, "a quality, which can only belong to the mind".[11] For convenience I will say that H4 describes very roughly the *performative* element of our concept of cause – according to Hume, of course.

2. HUMEAN NECESSITY

Characteristically, anti-Humeans tend to regard their primary issue with Hume to be necessity. Most of them claim that if *c* is the cause of *e* then there is a necessary connection between *c* and *e*. Yet often the disagreement is more verbal than real. To begin with, practically everybody agrees (and everybody should agree) with Hume that causal necessity is not logical necessity – just as H5 proclaims. It is true that for Hume there is only one concept of necessity, so that there is in his use of the word *necessity* no necessary connection at all between *c* and *e*. But these disagreements are all on the surface.

Hume could very well have insisted both that there is only *one* kind of necessity, in his use of the word, and agree that the word *necessary* can be used in *other* senses. He would have to add that all those senses do not have the requisite unity to constitute kinds or species of one master sense or type of necessity. Not to prejudge anything historical, let us speak of Hume*, rather than of the actual Hume, in this regard. Clearly, Hume* insists that the phrase 'causal necessity' has the word *necessity* not as a specifying adjective, i.e., an adjective that denotes a feature that determines a species of necessity. The word *causal* in the phrase 'causal necessity' is, for Hume*, a *semantical* adjective: it indicates the *sense* in which *necessity* is being used. (A useful paradigm is this: in the phrase 'extrinsic goodness' the adjective *extrinsic* does not denote a feature specifying a species of goodness, but only a different sense of the word *goodness*: as is well known, things that are said to be extrinsically good are not good – in the primary or intrinsic sense of *good*.)

Hume* also insists that the performative element (introduced by thesis H4) makes the concept of causal necessity irreducible to other concepts. That element is not conjoined to the generalization that formulates the constant conjunction, which is the cognitive content of a causal claim. The relationship between the two elements is not logical, so it is not subject to analysis. On the other hand, since we have some sensorylike consciousness of that performative element, we can attach a signal of it to the logical connective involved in the causal generalization. Typically, the connective of Humean generalizations is taken to be material implication, represented by Peano by the sign '$\supset$'. We can represent the Humean performative element of causality by '*'; thus, a Humean causal claim may be represented as having the connective '$\supset$*'. Thus, '$a \supset^* b$' analyzes for Hume* '*a* causes *b*'.

Hume* readily admits that we have a concept that can be expressed by '$\supset$*'. He is prepared to say that it and universal quantification are together the core of our idea of 'causal necessity'. He is curious to see it treated as a formal connective, in a formulation of the logic of causality. Since logical form must be freed from psychologism, we must not simply regard that '*' is attached to the statements a speaker claims to be causal. Obviously, a speaker may be inconsistent and may not pay attention to the sensorylike presentation of the performative element he is both engaged in and committed to when he makes a causal claim. We may treat the signal '*' as possessing some objective status by considering it as inheritable by what is implied by a Humean generalization possessing it. Thus, we have an idealized causal performative element and idealized connective $\supset^*$, which can be interpreted thus: it represents both an actual mental set of a person who makes a causal claim and the range of that mental set if he is logically consistent. It may be argued that in spite of our inconsistencies we have a deep-seated higher-order tendency to consistency. That is why we try to eliminate inconsistencies in our beliefs once we discover them. Thus, Hume* is very interested in the logic of causality. Knowing it will allow him to know how to maintain his 'causal' attitude through the whole range of what his 'causal' beliefs imply.[12]

My present point is this. Hume could have accepted the terminological observations presented above and then would have become our hypothetical Hume*. Hume* accepts the locutions 'causal necessity' and 'causally necessary' as indivisible expressions, in which the words *necessity* and *necessary* in their independent meanings he assigns them do not occur. Furthermore, Hume* accepts that there is something thinkable, i.e., an idea or concept, which those long, unanalyzable locutions express. Moreover, that concept is governed by rules that modal logicians may eventually be able to formulate correctly. Thus, Hume* does not disagree with a good deal of what anti-Humeans claim, allegedly in opposition to Hume.

Causality is not a logical relation. This is clear, thanks to Leibniz and Hume. Causality involves a necessary connection of some sort – this is not so clear, because we have to specify what we mean by 'necessary connection'. That this is so is clearly due to Hume. The main issue about causality is not, however, the necessity of the connection. (Hume* is prepared to accept a form of necessity.) The issue is *what exactly the causal connection is*. It may, or may not, be necessary. But its being necessary is something that accrues to it by virtue of being the connection it is, whatever it may be, and *this* is what we must determine.

3. THE HUMEAN PERFORMATIVE ELEMENT

We must set necessity aside. Hume's analysis of the causal connection between *c* and *e* is given by the theses H1–H4. On his view the causal connection consists of both a universal generalization and the performative element we described above. Patently, this will not do at all. The generalization is fine: it is an intersubjective element. On the other hand, the performative element is out of order: it is a subjective element. People react differently, so that some may acquire the custom (to use Hume's own word) of inferring the likes of *e* from the likes of *c*; but others may not become so accustomed. Thus, what is causally related for one person may not be so related for another.

The fundamental fact that *we live in one world with one causal structure*, which is the ground of our communication and cooperation, requires that our claims about causal connections be objective or at least deeply intersubjective. The only way for this to become plausible is for the Humean custom of connecting the likes of *e* with the likes of *c* to be deeply ingrained in the mind. That custom cannot be a mere policy defended on economy, laziness, or whatnot. A policy of this sort will have to presuppose something about the ways the mind works, which is precisely what we are saying. Hume himself hinted at this, but it was Kant who saw quite clearly what had to be done within the Humean framework to secure an intersubjective causal order.

We are not here, however, interested in modifying the Humean framework in order to establish the basis for the intersubjectivity of causal claims. We want, rather, to examine, not so much our claims about causality, but the causal connection itself, which we constantly claim to hold in our environment and in our own bodies. Here is a *fundamental datum*: the causal relation between a cause *c* and its effect *e* is not only intersubjective, but strongly assumed to be *objective*. Perhaps this objectivity is the hypostatization of what the mind, by a deep and inexorable law, places into certain generalizations. Perhaps, but even if this is so, when we think that *c* causes *e* we are not merely conscious, as Hume suggested, in a sensorylike way, vaguely, without conceptual articulation, of some attitude we create in ourselves both toward *c* and *e* and toward the likes of *c* and *e*. We think something definite that relates *c* and *e*, perhaps something that we throw on *c* and *e*, but something that appears, and we fully conceptualize it, as a very special relation linking *c* to *e*, something open to inspection by everybody, something conceived as belonging in the order of things, with its full objective status – whether we put it in by the working of our own mechanisms of experience or whether it is in the world by itself of its own. What is it that we perhaps put, and cannot help, putting into the world when we experience pieces of it?

4. HUMEAN GENERALIZATIONS

The only objective and fully conceptualized and, hence, cognitive element in a causal claim, according to both Hume and Kant, is nothing other than the corresponding universal generalization. This is what we must examine. Again, we continue to marginate the issue of the necessity of the causally relevant generalization.

All there is, objectively and cognitively, to the causation of *e* by *c* is, for Hume, the relevant generalization that the likes of *c* are contiguously followed by the likes of *e*. By Hume's own account such generalizations must have had instances in the past, even in one's own experience. Otherwise one cannot derive the requisite habit of inferring the likes of *e* from the likes of *c*.

As noted in H3 above, Hume speaks of events resembling the cause *c* and of events resembling the effect *e*. Mere resemblance is, unfortunately, an extremely unhelpful notion. Surely, they are all alike in being in space for instance. Everybody who has read Hume has undoubtedly remarked that it is exceedingly difficult to formulate the relevant Humean generalizations. Indeed, it is a commonplace to declare that in daily life we are completely sure that a certain event is the cause of another, even though we are unable to articulate a nonempty generalization that spans the whole universe without exceptions. For instance, we see Francesca kill Romeo by shooting him; we are immediately certain that her firing the gun in her hand at the time in question caused Romeo's death. But what is the relevant Humean similarity? We want a generalization that has instances in the past and has never faced a counterinstance. We all know the problem; I will not labor it.

A most fruitful suggestion has been made in recent times by Donald Davidson. He has proposed that when we claim that *c* causes *e* we are not claiming simply that events resembling *c* have been, and will be, followed by events resembling *e*. According to him:

D1. 'Event *A* caused event *B*' is true, just in case there are descriptions of *A* and *B* such that the sentence obtained by putting these descriptions for '*A*' and '*B*' in '*A* caused *B*' follows from a true causal law.[13]

Davidson uses this proposal to explain why we are unable to formulate the appropriate causal generalizations. This is of course a serious problem. But D1 does not help us at all to explain why we cannot formulate the Humean generalizations, which, as we well know, are not truly causal. Actually the Humean mystery is compounded rather than solved. It is not clear why we should in a given case have any confidence that there is a Humean or a causal generalization, whatsoever it may be, even though we cannot articulate it. Here the whole category would consist of existential generalizations of which we cannot know a single instance. Of course, sometimes we know that there is something, which we cannot specify. But in such cases we infer the existential claim from other premises. What sort of inference justifies us in the claim that there are generalizations we cannot formulate?

There are special reasons to be suspicious of the view that our language has the appropriate descriptions whose existence D1 proclaims. Laws of nature are typically formulated for closed systems. In daily life we never find truly closed systems. For instance, the laboratory that most successfully isolates certain physical experiments may receive some interfering cosmic radiation or it may be blown away by a tornado. The application of laws of nature to our daily experience requires the assumption, not that the system is really closed, but that the other conditions we are not taking into account do not matter: as, we say, that everything else remains equal. Thus, the causation of an event *e* by an event *c* typically involves a balance of circumstances and laws. Do we have, does our language have, the requisite Davidsonian descriptions for the singular causal claim to be true? I do not know. Yet I often *know* that a certain event *c* has caused an event *e*. The Davidsonian descriptions seem somewhat irrelevant.

Paradigm physical laws are not causal. They express functional relationships between, or among, parameters, without saying what is cause or effect. Consider 'The speed of light in a vacuum is constant and maximal', Newton's law of gravitation, Hooke's law, Boyle's law, and so on. They yield causal rules, which are subject to *ceteris paribus* clauses.

There is a small point worth pondering. It is not clear that we have a definite and useful concept of causal law. At least it is not clear that we have the conception Davidson proposes, namely, of a conjunction of the form:

D2. For every time t and for every event of type F occurring at t, there is exactly one event of type G occurring at t', very close to t, such that the former causes the latter, *and* for every event of type G occurring at t' there is exactly one event of type F occurring at t, which is the cause of the former event.[14]

I do not think that the cause-effect relationship is one-to-one. Hence, I do not claim that such causal laws back up my singular causal claims. Of course, we can trivialize such one-to-one laws: we can make them have as instances the events in a carefully chosen set of pairs of events, and put appropriate descriptions for them as substituends of '*F*' and '*G*'. But this is does not enrich our understanding of causality.

This is not the place to discuss Davidson's views.[15] We should note merely that, as he stresses, D2 is not an analysis of *cause*, and that the idea behind D1 is tame and safe. By D1 singular causal statements are made true by true causal laws (and minor premises about instantiation, of course). This is unobjectionable, *whatever* the causal connection may turn out to be. Our problem here, however, is not epistemological, but ontological. If we knew what the causal relation is we would be finished. We seek to understand what the causal connection is.

There is an important ontologico-epistemological datum hovering around Davidson's discussion, which we must not neglect, namely:

C.L*. We establish causal generalizations on the basis of singular causal instance pairs. We are more certain that an item *c* caused an item *e* than of any causal law under which *c* and *e* fall. We may be very sure, of course, that many laws of nature (formulated for closed systems), whether causal or not, impinge upon (or govern) the context in which *c* and *e* occur.

C.L*. tells us that causal generalizations are truly *causal* generalizations, to the effect that instances of certain types are indeed related by causality. This is not the case in Hume's view. For him the universal generalizations that make a singular causal claim true are not truly causal: they do *not* contain a genuine relation of causation, whether necessary or not; they contain no other 'causal' relation between the events of the appropriate types than constant conjunction in continguous space-time. I am here ignoring the performative, noncognitive element signaled by '*' in Section 2 above for the reasons given in Section 3.

5. CAUSAL GENERALIZATIONS AND HUME'S *ABSTRACT*

Humean generalizations are not truly causal: they do not include a special relation of causation. They are, according to Hume's thesis H3 above, formulations of universal coincidence. This is objectively the whole of causality in the Humean view. Is this correct? Let us consider a typical example examined by Hume himself:

Here is a billiard-ball [let us call it *ball B*] lying on the table, and another ball [let us call it *ball A*] moving towards it with rapidity. They strike; and the ball [*B*], which was formerly at rest, now acquires a motion. *This is as perfect an instance of the relation of cause and effect* as any which we know, either by sensation or reflection. Let us therefore examine it. 'Tis evident, that the two balls touched one another before *the motion was communicated*, and there was no interval betwixt the shock and the motion. *Contiguity* in time and place . . . . *Priority* in time . . . . But this is not all. Let us try *any other balls* of the same kind in like situations, and we shall always find, that *the impulse of the one produces motion in the other*. Here therefore is a *third* circumstance, *viz*. that of a *constant conjunction* betwixt the cause and effect. Every object like the cause, *produces* always some object like the effect. Beyond these three circumstances . . . I can discover nothing in this cause. The first ball is in motion; touches the second; immediately the second is in motion: and *when I try the experiment* with the same or like balls, in the same or like circumstances, I find, that upon the motion and touch of the one ball, motion always follows in the other. In whatever shape I turn this matter, and however I examine it, I can find nothing farther.[16] *(Quoted by permission of the Oxford University Press.)*

This is a remarkable summary of Hume's discussion of causality in the *Treatise*. It makes it palpable, so to speak, how there is no logical necessity in the motion of ball *A* being followed by the motion of ball *B*. But let us examine the situation Hume describes in slow motion, so that we can savor its main points.

Point 1. Hume agrees with most of us in that an ordinary person's experience in an ordinary billiard-ball game, involving ordinary billiard balls and ordinary billiard tables, does present a perfect example of causation. Of course, we may not be able to tell whether the billiard balls or the table are of the ordinary kinds. But if they are, whether we know it or not, we are perceiving a perfect case of causation. We may not be able to assert this with confidence once we develop doubts about the ordinariness of the items involved. It does not matter: if they are of the ordinary kinds, we *are* seeing a perfect case of causation.

Point 2. Undoubtedly, the conditions enshrined in Hume's theses H1 and H2 are present.

Point 3. The condition of constant conjunction is not perceived, of course. Hume introduces it in the most revealing way. We must go very slowly on this. Instead of referring us to his past experiences of billiard balls, where the full resemblance to ball *A* and ball *B* should help, Hume does something extraordinary. He goes on to "try any other balls of the same kind in like situation". He also tries the same balls. In all, he turns to future cases. His claim that the collision between ball *A* and ball *B* is a perfect example of causality is not supported only by past cases. He needs them for him to *be able* psychologically to claim that the *A-B* collision is a perfect case of causation. But when he goes on to *justify* his claim he goes into the future. This is fine, since the generalization involved in the claim of causation must be a universal one, ranging over the whole of time. Yet there is something perplexing here. How does Hume expect to verify conclusively the appropriate generalization by considering a few, or many, other cases? The future time surely goes *much* beyond the time at which he ends his experiments. Since the generalization is not logically necessary, on his official view the confidence he shows is only an expression of his habit of inferring motion in the likes of *B* from motion in the likes of *A*.

Point 4. Hume is sure that he has found all the causality he wishes to establish. He says most deliberately: "and we *shall always* find that the impulse of the one [ball] *produces* motion in the other [ball]". Hume has said many times that the verb *produce* is just another synonym for *causes.*[17] Hume is sure that the striking of one ball by another causes, produces, the latter's motion, every time one ball strikes the other in a similar experiment as the one involving ball *A* and ball *B*.

The point is that *Hume establishes the causal generalization by induction from the singular causations*! This is precisely the most fundamental datum C.L* discussed above. It is in fact a brilliant, perfect application of datum C.L* as any which we may come to know.

Hume is so definite about this, that he underscores the point by formulating the appropriate true causal generalization explicitly: "Every object like the cause, *produces* always some object like the effect".

Point 5. Hume's discussion reveals an important principle, which we may christen *the piggybackness of causality*:

C.G. If an item c causes an item e, then there are properties ϕ-ness and ψ-ness, and circumstances Z such that c is ϕ, e is ψ, c is in Z, and whatever item of the same category as c is ϕ and in circumstances Z causes an item which is ψ and of the same category as e.

This important principle C.G. is reminiscent of Davidson's proposal D1. It is also like D1 in that it does not provide an analysis of the causal relation, but only a connection between a singular causal proposition and a general causal proposition. It differs from Davidson's proposal in that it makes no claims about the existence of, or the richness of any existing, language. It also differs from D1 in that the causal generalization implied by a given singular causal proposition need not be a causal law – whatever this may be.

Point 6. Up to this point Hume's text is clear. Yet he repeats the discussion point by point, except for one thing. He makes a list of what he sees in each experiment, and now he lists the contiguous succession, the contiguous location, and the constant conjunction of the events composing each experiment.

He does not mention now that the striking of one ball by another produces the motion of the one. What has happened? Here we have Hume's official doctrine minus the performative element. Clearly, Hume is mistaken in saying that regardless of how he examines the situation he finds nothing farther. For one thing, there is the missing performative element. This is required for him to justify his claim that the striking produces (causes) the motion in the struck ball. He cannot know the generalization to hold for the whole of the future time – not to mention the remote past. His confidence that the situation provides a perfect example of causation has to be grounded on something accessible to him. On his official view that something can be only the performative element. Given our objections to the performative element, we must feel unhappy with Hume's impatience at not examining the matter more deeply.

Hume must, nevertheless, be praised for applying crucial datum C.L* above. The passage shows very perspicuously that Hume is correct in claiming that he has in the striking of ball *B* by ball *A* a perfect example of causation. Let us attempt to find out why.

6. HUMEAN GENERALIZATIONS WITHOUT CAUSALITY

An ordinary person would say that in Hume's billiard-ball example there is causation of the motion in ball *B* by the collision of this ball with ball *A*. And Hume agrees. But the official Hume claims that *there*, on the billiard table, is no other relation than contiguous succession. The ordinary person would say (with the actual Hume who wrote the passage above) that the collision *causes, produces* the motion in ball *B*. Is there any relation of production over and above contiguous succession? This is the question. It is not easy to answer. But one thing is clear. If the ordinary person is right, then there is something in Hume's billiard-ball example which would not be present in a similar example in which causation did not supervene on contiguous succession. This is indeed clear. And we must immediately proceed to a *Gedankenexperiment* to test the separability of that alleged causal relation from succession.

The example is, as Hume tells us, a perfect example of causality. Thus we must attempt to remove causality and leave the contiguous successive conjunction of collision and motion. Furthermore, we must guard our rear. Since the official Humean doctrine includes a universal constant conjunction, we must not tamper with this. Obviously, if this goes, so goes causation for the Humean. Thus,

> *Desiderata*. The experiment must maintain: (i) the collision of ball *A* with ball *B*; (ii) the motion in ball *B*; (iii) the immediate succession of the collision and the motion in *B*; (iv) the truth of the universal constant contiguous conjunction of similar collisions and similar motions. But (v) it must not maintain the causation of the motion in ball *B* by the collision of the two balls.

Once the desiderata are spelled out in detail it is embarrassingly easy to provide an example that satisfies them. Here is one:

> G.-C. Here is billiard ball *B* lying on the table, and ball *A* is moving toward it with rapidity. There is a mechanism *M* under the table such that when ball *A* reaches ball *B*, mechanism *M* stops ball *A* cold and prevents ball *B* from moving. There is also another mechanism *M′* such that it is timed by a certain clock to release the hold of *M* on ball *B* and set it in motion with a certain speed. It happens that *M′* sets ball *B* in motion at the time ball *A* reaches *B* and the speed is the same as the one *B* would have had, had it been caused to move by ball *A*.

Clearly all five desiderata are fulfilled. The example is even better than what Hume's discussion required. We have added a condition on the motion on ball *B* that Hume did not consider. Obviously, the example can be enriched so as to meet an improved example of Hume's, one that brings in relevant features of mechanics that Hume ignored, e.g., the direction of the motions and the velocities and the masses of the balls. A sophisticated Humean philosopher may wish to bring them in as requirements on the 'causal' generalization. We may suppose them all included in G.-C.

The point of example G.-C is that Humean generalizations of constant conjunctions may hold not only because they describe universal coincidences, but also because they may be partly causally true and partly coincidentally true. Evidently, the performative element of Hume's analysis of causality may be present hovering around G.-C. A person who knows the mixed truth conditions of a Humean generalization has no reason for not maintaining his inferential habits to predict an item like the effect on the presence of an item like the cause.

In example G.-C we hypothesize that there are two causal mechanisms *M* and *M′* which bring about the coincidence of two causal paths. We have removed the causal connection between the motion of ball *A* and the motion of ball *B* by linking them to two different causal chains. Mechanism *M* literally *absorbs* the motion of ball *A*, and mechanism *M'* causes ball *B* to move. We have, a Humean may remark, introduced new sequences of contiguous successions of changes. He will add that there will he Humean generalizations of which they are instances. Perhaps, although it will not be easy, once again, to formulate precisely those generalizations. Yet it is important to observe that causality can be drained from those sequences of contiguous successions with changes by reapplying the same technique of example G.-C. We can introduce for mechanism *M* a pair of mechanisms *N* and *N'* such that *N* deviates the causality of *M*, and *N'* reintroduces the effect of *M*. Similarly for mechanism *M'*. This may introduce new contiguous 'causal' successions of changes; but these in their turn can be drained of their causal factor.

7. CAUSITY, MOTION, AND ENERGY

What, then, is present in the causal relation between the motion of ball *A* and the motion of ball *B*? It is extremely hard to say. It seems, as Hume unwittingly illustrated in the quotation analyzed above, that we cannot but describe the causal relation except in terms that look question-begging.

We can say, with Hume, that the collision *produced* the motion in ball *B*, that ball *A communicated* its motion to ball *B*. If that is really what counts, then we must simply acknowledge that when an item *c* causes an item *e*, then something *passes* from *c* to *e*, or something is *destroyed* in *c* and a replica is *created* in *e*. This idea of causal communication of something deserves careful reflection.

In Hume's example we can say that (some of) the motion of ball *A* literally transferred to ball *B*. But we cannot identify in general the transfer of motion with causation. Yet the transfer of motion is precisely the aspect of the situation that Hume did not perceive well enough. In our example G.-C there is no transfer of motion from ball *A* to ball *B*.

There is, however, something perplexing about a transfer of motion. This suggests that motion is being hypostatized.[18] The view that the motion of ball *A* is destroyed (or part of it, if it bounces from ball *B* at a lower speed) and that the motion of ball *B* is created at the appropriate subsequent instant seems better. In any case something of this sort seems to be clearly involved in Hume's billiard-ball example. We shall for convenience continue to speak of the *transfer of motion*, allowing for either a literal transfer or a destruction of motion in the cause object and a corresponding creation in the effect object.[19]

In our example G.-C nothing leaves ball *A* to go to ball *B*. Mechanism *M* absorbed the motion of ball *A*. But it is not required that anything in mechanism M receive that motion. In any case, we know that there are causes that do not, at least in appearance, involve motion. But perhaps every causation involves a transfer of motion. This is an empirical matter that we cannot decide by merely reflecting on the general nature of causality. The crucial thing is that in causality there is a transfer of something from the cause to the effect. For convenience, neutrally, so as to beg no empirical questions, let us call *causity* that which is characteristically transferred from a cause to its effect. Again, we shall speak of *transfer of causity* with the same generality we decided upon in the preceding paragraph regarding the transfer of motion: we allow that there may not be a literal transfer across time of some selfsame item, but only an appropriate matching of a vanishing of causity at a given space-time and a contiguous creation of causity at an adjacent space-time.

At this juncture we must leave it open whether causity is motion or momentum, or whether causity is something generic that allows of several different specifications. Empirical considerations suggest that motion itself is not causity, but that it just happens that when motion is transferred, as in Hume's billiard-ball example, there is also a transference of causity or of force as Leibniz says.[20] In particular we must leave it open whether causity is exactly what scientists nowadays call *energy*. If this is the case, we must hold that the identity of causity with energy is a *contingent identity*. Definitely, what precisely causity is is a topic for scientific investigation. At our present abstract level of philosophical reflection causation reveals itself as much more than universal coincidence of properties in contiguous succession. Causation is a transfer of something from cause to effect. Philosophical reflection can say nothing concrete. It must formulate the most basic ontological and epistemological principles. But it cannot substitute for scientific investigation: it must, on the other hand, provide a foundation for scientific investigation. Perhaps what science nowadays calls energy is all there is to causity. But perhaps there are still unknown forms of energy, i.e., of causity. And perhaps there are further forms of causity that should not be called energy because of their anomalous or very different properties. Perhaps psychophysical interaction involves a form of causity which is not physical energy of the well-known types. All these questions belong, not to the philosophical theory of causality, but to scientific theorizing, even if the answers require a broadening of current scientific theorizing and its concepts and techniques.

8. CAUSITY, EXISTENCE, AND PARTICULARS

We have established that causity is not a property, and we have claimed that it is a fundamental ontological parameter of the world structure together with space and time. To further clarify the ontological status of causity let us note that *causity is not existence*. Causity is what is transferred from the causes to their effects. A little reflection reveals that that cannot be existence. In the case of particular events, which embody their respective places and times, a transfer of existence from the cause event to the effect event amounts to mere succession. The cause occurs and then the effect occurs in a contiguous place. In fact, we can gain more insight into Hume's view of causality by noticing that it amounts to the identification of causity with existence.

Existence is not a quantity. It is, like universals, exactly the same everywhere it occurs, and it allows of no degrees. Of course, one may need to distinguish degrees of reality, but each degree, whether we call it existence or not, is the whole *unitary* selfsame entity at each of its *diverse* occurrences. Existence behaves in our apprehensions of it just like a property or a *universal*. It is, on the other hand, a formal property; it differentiates very little: it is the boundary of empirical discourse.

Causity is a quantity and must be measurable, as we shall discuss subsequently. Hence, here is another argument for the distinction between causity (and energy) from existence.

Causity is a quantity and is in principle measurable. But it need not appear in natural units, in grains of causity (or energy), so to speak. Of course, it may do so either wherever it appears, or just in some particular manifestations. But amounts of quantity are not objects. The transfer of energy from a cause event to an effect event is not the transfer of a particular which is the transferred causity. Notwithstanding, the transfer of causity from one object to another may ride, although it need not, on the transfer of a piece of the cause object to the effect object. Because this is an instance of causity transfer we can perceive gross instances of causation. Yet we must insist that the transferred causity is not identical with the transferred object – just as before we said that transferred motion is not identical with transferred causity. Causity goes often from object to object riding on the backs of transferred pieces of matter. Causity remains an abstract structure, like space, and time, and, like them, quantifiable.[21] Yet, as in the case of transfer of matter and motion, it may be perceptible.

9. THE DEPENDENCY OF CAUSITY ON PROPERTIES

In Section 5 we already noted that causation rides piggyback on some properties of causally related items. We formulated principle C.G, which grounds causal generalizations. Yet it is important to stress that causity is not identical with the properties involved in causal generalizations. Causity is that pervasive element that transfers from a cause to its effects. Certainly causity cannot be identical with the properties of objects involved in changes. These properties change, whereas causity abides through the time of change. For any property *P* of an object *o*, if *o* can become not-*P*, the causity transferred across the time of change can, clearly, be identical neither with *P*-ness nor with not-*P*-ness. In the case of so-called substantial change, when an object *o* is destroyed in a change from being *P* to being not-*P*, whatever the subject of the change may be, it is still the case that neither of these properties involved in the change can be identical with causity.

Suppose that all changes involve an underlying immutable Aristotelian substrate. This substrate may be Newtonian or Einsteinian absolute space. Then that substrate cannot be causity. Causity transfers from one space-time position to another when a caused change takes place. The substrate does not transfer.

In brief, then, *causity is a pervasive aspect of the universe which must be distinguished from the properties* (*qualities or relations*) *of objects*. It underlies, indeed constitutes all caused changes. It is also multifarious. But causity is itself imperceptible. Like the other fundamental ontological parameters of the universe (time and space), causity is a formal structure to be filled in by the contents of the world. Thus, in order to perceive causation, i.e., a transfer of causity, we must perceive the succession of other properties in contiguous space-time positions.[22]

At a very high level of abstraction one can conceive of causity standing alone in reality, with all its connections to the properties of objects wholly severed. One can conceive of causity transferring across time, regardless of any regular succession of the properties of objects. [23] That one can conceive of property regularity going one way and causity transfer another way is evidence of the mutual distinctness of properties and causity. But it seems to me that what one thus conceives is not a viable world. In such a conception, causity is an unfathomable aspect of reality wholly transcendent to our experience. For concreteness, since energy is a form of causity, we are conceiving of a world in which there is transfer of energy from object to object, but that transfer cannot be connected with the regularities we find in experience, e.g., the regular patterns of changes in the volume, the pressures and the movements of objects. In short, we would not have the laws of energy transfer we have. Furthermore, there would not be any other laws to replace the ones we have. Energy would not be governed by laws relating it to the properties of objects. The postulation of energy would be entirely gratuitous.

Causity must be related to the properties and relations of objects. It must be so related that its presence can be ascertained. But it must be related to properties for the universe to gain a thoroughgoing orderliness. I would like to argue that a thoroughgoing orderliness of the universe is required for it to have life, or at least consciousness, and that even a higher level of orderliness is required for the universe to have agents, i.e., beings who think and have the power of deliberation and volition. But there is no space for this here.

10. OBJECTS, CAUSITY, AND FINITUDE: THE MEASUREMENT OF CAUSITY

We tend to speak of an event or change as causing another event or change. This is the tendency in the Humean tradition. In terms of causity (or energy) we must distinguish the objects that stand in causal relations from the changes or events of those objects. The changes are the ones that are productively involved in the transfer of causity or energy. This is precisely the underlying point in the thermodynamic distinction between the work an object performs and the heat it receives or gives out, on the one hand, and the potential energy of an object, on the other hand. It is easier to figure out all the transformations of energy in terms of the amounts assigned to objects and fields than it is by just concentrating on the differential amounts pertaining to changes. But this is not conceptual: it is a matter of computation.

Similarly, in our more general talk about causation we will find it easier to speak of the causity of objects and of its decrease or increase, rather than of the active causity that is transferred from change to change. We must distinguish, in any case, between activated and nonactiavated causity, the former being involved in changes.

We have spoken of amounts of causity. This is justified a posteriori by the fact that energy appears to be a specification of causity. Nevertheless, there is a more general consideration that suggests that causity comes in amounts. Often a cause has several effects. Hence, the causity in the cause divides into the amounts going into the several effects.

Causation is measurable. The measure of causation is just the measure of the causity that transfers from cause to effect. We must dwell upon this in order to formulate some of the most fundamental principles involved in the quantization of causity. These principles are foundations for the measurability of energy, and for the measurability of other forms, if any, of causity.

The most elementary principle required for the measurement of causity connects directly with the ontologico-epistemological dependency principle C.G of Section 5. The following principle provides a foundation for C.G:

C.M.1. *Causity dependency*. Sets of properties, not the individual properties of an object, or field, are assigned (by nature, of course) amounts of causity.

C.M.2. *The possibility of epiphenomenalisms*. The amount assigned (by nature) to a set of properties may be zero.

A set *S* of properties is epiphenomenal for an object, if and only if the properties of *S* are causally inefficacious. For that, given our previous results, it is necessary that the set *S* be assigned by nature zero causity. I hasten to stress that nothing so far is evidence for, or against, the view that purely psychological properties are epiphenomenal. Which properties are epiphenomenal is an empirical matter of the greatest importance. It cannot be decided by philosophical analysis.

By principle C.M.1 sets of properties, rather than individual properties, are assigned causity, in order for the measurement of causity to be manageable. For causity (or energy) to be measurable it must appear in finite amounts. But then not all properties of an object can be assigned nonzero amounts of causity, unless, of course, these amounts form a convergent series. The reason is that every object has an infinity of properties. For instance, for any property you consider, say, ϕ, there is an infinity of logically equivalent properties generated by the iteration of the schema 'ϕ or (φ and ψ)', where the place of 'ψ' can be occupied by 'ϕ' itself. Thus, C.M.1 lays it down that each object has causity assigned to it by nature assigning amounts of causity to certain sets of the properties it exemplifies.

The partition of the properties of an object *o* into sets of logically equivalent properties does not guarantee the finitude of the amount of causity assigned to *o*. Those sets may be infinite in number. Indeed, if space is infinitely divisible an object has an infinity of spatial properties: distances to the infinitesimal sectors in which a nearby volume can be divided. We need another assumption to secure the needed finitude of the causity of an object.

We mention above that one useful assumption would be that each of the properties of an object *o* be assigned amounts of causity in such a way that *all* the assignments can be arranged in a convergent series. This is, however, a very drastic assumption. It would require that all the properties be taken together somehow, so as to guarantee the convergence of the assignments. Moreover, since the order of the terms of a convergent series is crucial to its convergence, all the properties, or equivalent sets thereof, have an absolute ontological order. Yet there is no way we can figure it out. Naturally, I cannot prove that the universe is not really like that. I, like Leibniz, believe in the thoroughgoing unity and deep orderliness of the universe. Yet I cannot find evidence, or some general reasons, to suppose that the universe is *convergently unitary and orderly*, and that its properties have a fundamental ontological order for causity assignments.

In the case of some particular properties we may certainly find that the assignment of causity to families of them is the sum of a convergent series. If the space occupied by an object *o* is infinitely divisible, then the amount *C* of causity assigned to *o* connected with its space is also infinitely divisible, since every part of *o* will have some causity. Then the assignments of causity to all the division points converge to *C*. We have in such cases continuous functions measuring the causity of an object along certain families of properties. Still, the assignments to independent families of properties seem to be independent of each other.

It seems, therefore, in agreement with common facts of experience, that at least *some* pairs of independent sets of logically equivalent properties are assigned (by nature) their respective amounts of causity independently of each other. Nature appears, however, more economical in that it assigns causity, not to sets of logically equivalent properties, but to larger sets of properties, which may be called *causally equivalent*. This is *perhaps* illustrated in the case of the normally perceived colors of objects and their powers to reflect light of different wavelengths. There is some causal equivalence between sets of color shades and sets of properties involved in reflection of certain lengths of light waves. But the extent of the causal equivalence is far crom clear. The psychological effects in color perception are not fully understood.

Perhaps at bottom there is only one family of sets of properties that are assigned causity, the other families, thus, being epiphenomenal. This is the general assumption embedded in the views of those philosophers who hold that the only causality there is in the universe is the causality of the ultimate particles physics postulates. That might be true. At any rate, from a general discussion of the nature of causality we cannot establish reductionist programs of any sort, much less those which require the most extensive empirical investigations. Nor can we establish their denials either. To sum up, causity is governed by the following fundamental principles:

C.M.3. *Causity finitude*. The total amount of causity an object possesses and the amount of causity assigned to a set of properties are both finite.

C.M.4. *Causal extension*. If a set S of properties of an object o is assigned an amount C of causity, then S is closed under logical equivalence. That is, if ϕ is in S and ϕ is logically equivalent to φ, then φ is in S.

C.M.5. *Additivity of causity*. If an object o is made up of objects a and b, then the causity of o = the causity of a + the causity of b.

For convenience we shall speak of *representative causal subsets* of sets of equivalent properties. Given a set S of causally equivalent properties, we can consider any subset σ of S whose members are pairwise not logically equivalent. To the extent that nature makes no economies in its assignment of causity, a representative subset is a unit set. Clearly, when a representative subset is not a unit set, then we have a basic law of nature, namely, one that establishes the causal equivalence of the members of the representative subset. Such laws are not causal laws, properly speaking, because they do not give specific conditions for measuring the amount of causity that transfers from cause to effect. Those laws merely say that causally speaking the members of the representative set can be used for measuring the transfer of causity. We shall call them *laws of causal equivalence*, rather than causal laws. Now, we have the principle of:

C.M.6. *Causal economy*. The more economical nature is in its distribution of causity, the more laws of causal equivalence order the universe. There is some causal economy in the universe whenever there is a representative causal subset σ with n members and $n > 0$.

Undoubtedly this topic must be pursued. There are many other causal principles, but we must move on.

11. CONSERVATION OF CAUSITY AND OF ENERGY

Let us make another *Gedankenexperiment*:

C. The universe is in a certain state S such that only one section of it is filled with circumstances C and there is in C one object a moving at a time t. Let us call c the event of a's motion. Suppose that at time t', an interval consecutive with t, c causes event e in object b, everything else remaining at t' as it is in S at t.

By the preceding principles, causity (energy, momentum, heat, and so on) transfers from *a* to *b*, and event *c* transfers causity to event *e*. We must now ask:

(1) Must *c* transfer all its causity to *e*?
(2) Must all the causity that *a* loses be gained by *e*?
(3) Can *e* have more causity than it receives from *c*?
(4) Can *b* have more causity than *a* loses?
(5) Can the universe have more causity at t' than it has at *t*?
(6) Can the universe have less causity at t' than at *t*?
(7) Must the universe have the same amount of causity at t' as it has at *t*?

Evidently, there is nothing in our analysis of causation so far that can provide an answer to any of these questions. All we know for certain at this juncture is that object *a* loses some of its causity, and that event *c* at *t* loses causity, too. We can argue that event *c* just occurs at *t*, so that the motion (if any) of *a* after *t* is another event in the history of *a*. In that case *c* disappears altogether and so does its causity. But some causity will be replicated in the new motion of *a*.

As far as our analysis has reached, it is permissible that the universe as a whole may have less causity at t' than it has at time *t*. It is, so far, not required that the causity in the world remain constant. In a fully causally determined universe, no event is uncaused, hence there is no injection of causity which does not come from previous causity. *In a deterministic universe the amount of causity will not increase*. Nevertheless, so far as our analysis of causation has proceeded, a fully deterministic universe might very well lose some causity in the transition from causes to effects. The total effect state in circumstances *C* must have causity issuing from its causes; but nothing requires, so far, that it receive all the causity that the total cause state has. The universe, then, could have causal leaks: some amounts of causity could just vanish without reappearing, or without being replicated, in any effects. Indeed, the universe could, so far, have causal leaks even in regions where no effects are occurring. Causity could become, so to speak, rusty, untransferrable, and would just vanish. Those causal leaks might occur in accordance with some decreasing ratio; then the causity in the universe would always be approaching some limit.

Could the universe lose all of its causity and disappear? Here we have two questions. That the universe might lose all of its causity is not precluded by anything so far discussed. But disappearance is something else. The disappearance of causation is not the same as the disappearance of the universe. The universe might go on, but there would not be causation in it. Since we have assumed, for the moment, a fully causally determined universe, the universe would not be gaining any new, untransferred causity.

Now, as far as mere causation goes, the universe could have a history of steadily diminishing total causity. But the whole point of postulating causity is to introduce an ontological foundation for the thoroughgoing orderliness and deep-reaching unity of the universe. There is also the ontologico-epistemological requirement that that order and unity be at least partially accessible to our systematic investigation. We must, therefore, assume that the thoroughgoing orderliness and deep-reaching unity of the universe is not beyond man's capacity of comprehension – at least up to a point. Or, alternatively, we may suppose that the universe presents us with a simplified orderliness and unity suitable for our cognitive capacities. That in-principle understandability of the order and unity of the universe includes, thus, a *principle of simplicity*. Even if this principle lacked the full ontologico-epistemological support just suggested, it certainly has methodological validity. We must, for epistemic economy, suppose the universe to be relatively simple in our earlier investigations, keeping ourselves in readiness to move on through small degrees of complication – as small, but not less small, than the data at hand require. We want both conceptual and computational simplicity. A principle of simplicity applied to the measurement of causity immediately leads to:

C.M.7. *Conservation of total causity*. If the universe is fully deterministic, the total amount of causity in it is constant throughout all changes.

C.M.8. *Conservation of partial causity*. Some sectors of the universe are causally closed in that the total causity in them remains constant throughout changes in them. In some sectors causity remains approximately constant.

These principles allow the formulation of equations with functions measuring different assignments of causity to two different sets of properties, one in the cause and the other in the effect. Naturally, other conservation principles are forthcoming. Let us sum up:

(1) causity can be introduced into the universe by the appearance in it of uncaused events or objects, which become causes;

(2) the conservation of causity (or of energy) is not required for the universe to be fully causally determined: it needs simply that causity not be increased;

(3) the universe can lose all its causity without disappearing;

(4) a universe without causity can still continue changing: but then no change has causity: thus all changes could be uncaused and ineffectual;

(5) principles of simplicity about the thoroughgoing order of the universe lead to principles of causity conservation in the presence of determinism.[24]

12. HUMEAN VERSUS LEIBNIZIAN UNIVERSES

A universe without causity may still change, we said. No reason is there to suppose that a changing, though uncausal universe, may not have regular changes. Let us suppose that the changes are all regular, and have no exception across time. In short, let us conceive of a universe that has no causality but possesses as many Humean constant conjunctions as is possible. That would be a *truly Humean universe*. What does that universe miss by missing causation? This is the converse of the question we asked earlier about Hume's billiard-ball example, but transferred to the universe as a whole. We must discuss it in order to see the issue from the other side.

The changes in the truly Humean universe have no causes and have no effects. No transfer of causity (or energy) is involved in the succession of changes. Because of the dependency of causity, a truly Humean universe lacks the regularities that allow us to determine that causity is being transferred from cause to effect. These are regularities that unify perceptions in a higher, abstract order. As in the case of energy tranfer, we have abstract regularities unified within a system of equations, but it may very well be that no particular sets of values of the variables of those equations, which allow us to measure energy, have been conjoined more than once in the whole history of the world. This point is fatal for the phenomenalistic view of causation that Hume adopted. But it applies also to the Humean view of causation embedded in many nonphenomenalistic accounts of physical objects.

A Leibnizian universe, by contrast, has the coincidental regularities of perceptible properties, those of nonperceptible properties, and the abstract unperceived regularities of systematic transfer of causity (e.g., energy) from antecedent cause to contiguous consequent effect. These are, again, regularities in that they conform to certain systems of differential equations, but need not involve Humean constant conjunctions.

The difference, then, between a truly Humean and a Leibnizian universe lies both in the postulation of causity (or energy, and so on) and in the deeply thoroughgoing unity and orderliness of the Leibnizian universe, as contrasted with the superficial orderliness of the truly Humean universe. In the Leibnizian universe the Humean exceptionless constant conjunctions are undergirded by a network of higher-order principles of order. These are the laws of nature, including the laws of causal equivalence, all of which govern the transfer of the postulated causity.

The postulation of causity is a hurdle. One feels that given that causity is *per se* imperceptible, all one has to do to understand the complex order of the world is to know as many constant conjunctions as is possible and to learn the unification that the intricate laws of nature and of causal equivalence bring about. Why postulate causity? Undoubtedly, the postulation of causity has heuristic value. This is so, especially if it is postulated with other laws in addition to the basic ones formulated above.

I really do not know whether causity can be eliminated. A type of phenomenalism about causity, which analyzes causity in terms of the values of the equations for causity transfer, seems to me to be flat and unilluminating. *After* we have acquired a concept of energy as governed by a long list of equations, we can propose the elimination of the concept of energy as an aspect of reality, and propose regarding the regularities introduced by the equations as an ultimate brute fact. This leaves, for one thing, no unifying thread. For another, that can look minimally plausible only after the discovery of the laws, not before. When we are investigating the regularities to be formulated in our equations we need some unifying concept, and that is precisely what the ideas of causation and energy provide.

Patently, the preceding considerations do not compose a conclusive argument against a phenomenalism of energy or causity. Perhaps hypostatizations that have enormous heuristic value can be treated just as heuristic tools, both to be used and to be discarded once we have found our equations. Yet we should make sure that they are no longer needed, that we have completely exhausted their heuristic value. Clearly, that reduction can be an ideal of reason (to use Kant's expression), but it represents a Ptolemaic point of view, which leaves us at the mercy of Copernican principles needed to describe and explain the world as we find it in the ongoing process of living in it and theorizing in it about it.

On the other hand, I do not know that a phenomenalistic program for the elimination of causity will succeed. The program requires so many and such complex empirical presuppositions that only the optimist breed can feel unrestrictedly confident about its success.

Now, if we could cast away the postulation of causity and feel happy with the abstract order that the complex equatons of science deliver, we could improve our Humean bias against imperceptible entitles or aspects of reality. Truly Humean worlds are clearly less orderly than the world present physics describes. But a *super-Humean* world which adopts the deep and thoroughgoing orderliness of the Leibnizian world, but insists on a phenomenalist reduction of causity and energy to the order of the nomological equations of sciences, nominalistically understood, may perhaps gather as much order as a Leibnizian world. Of course, there are *sub-Leibnizian* worlds, in which the phenomenalist reduction of causity and energy and the like has been produced successfully. These can be ordered so that they can approach a super-Humean world. The theory of that convergence should be most illuminating. Any interested metaphysical logician?[25]

Indiana University

NOTES

[1] I learned the expressions 'concept-empiricism' and 'judgment-empiricism' from Wilfrid Sellars.

[2] All the page references to Hume will be to David Hume, *A Treatise of Human Nature*, ed. by L. A. Selby-Bigge, rev. by P. H. Nidditch, Oxford University Press, London, 1978. Among the useful things Nidditch has done is to reincorporate Hume's *Abstract of the Treatise* (pp. 643–662).

[3] *Treatise*, 1.3.2, 14.

[4] *Ibid.*

[5] *Treatise* 1.3.14, p. 172.

[6] *Ibid.*

[7] *Ibid.*, p. 171.

[8] *Ibid.*, pp. 171–173.

[9] *Ibid.*, p. 164. James Anderson has developed an ingenious argument in pursuit of the sensory impression that on a Humean concept-empiricistic view provides the crucial, specific content of our idea of cause. See his "The Truth in Voluntarism" (forthcoming).

[10] *Enquiries Concerning Human Understanding and Concerning the Principles of Morals*, ed. by L. A. Selby-Bigge, rev. by P. H. Nidditch, Oxford University Press, London, 1975.

[11] *Treatise* 1.3.14, p. 168.

[12] The logic of $\supset^*$: it distributes through conjunction: $a \supset^* (b\&c)$ implies, and is implied by, $(a \supset^* b)$ & $(a \supset^* c)$. Humean necessity amounts to a generalization over times of the form: $\forall t\ (At \supset^* Bt)$, i.e., *A*'s have caused and will cause *B*'s. Humean generalizations conjoin: $\forall t\ (At \supset^* Bt)$ & $\forall t\ (Pt \supset^* Qt)$ implies $\forall t\ [(AP)t \supset^* (BQ)t]$. We can define Humean, rather Hume*an, necessity as follows:

> $N^*(p) =_{\text{Df}}$. There is a Humean generalization *G* such that *G* implies logically that *p*.

This yields some obvious standard theorems. Depending on how one conceives logical implication, one may find an S4-structure, or something else. The topic is not entirely lacking in interest.

[13] Donald Davidson, 'Actions, Reasons, and Causes', *Journal of Philosophy* **60** (1963), 685–700, p. 698, note 7. This paper has been justifiedly celebrated and has been included in many anthologies. The same principle is developed in more detail in his 'Causal Relations', *Journal of Philosophy* **64** (1967), 691–703, but the succinctness of the above quotation is admirable.

[14] Davidson, 'Causal Relations', p. 699.

[15] I plan to do so in a sequel, 'Events', to this essay. A third paper will be 'Causal Necessity'. I originally conceived the three studies as parts of one essay in honor of Richard Taylor, who has thought deeply about causality. But the first part turned out to be long enough.

[16] *Abstract*, p. 649ff., some emphasis added.

[17] See, e.g., *Treatise* 1.3.14, p.157.

[18] Compare Leibniz's view on motion: "There is nothing real in motion itself except that momentaneous state which must consist of a force striving toward change" (*Specimen Dynamicum*, written in 1695, in *Philosophical Papers and Letters*, trans. and ed. by Leroy E. Loemker, 2nd ed., Reidel, Dordrecht, 1969 pp. 435–452, p. 436). All citations are from Loemker's translation.

[19] Compare Leibniz's view of transfer of motion, place, and so on: "it being impossible that the same individual accident should be in two subjects or pass from one subject to another" (*ibid., The Controversy Between Leibniz and Clarke* 9: 'Leibniz's Fifth Paper' 47, pp. 674–721, p. 704).

[20] See note 19; *cf*: "the same force and vigor remains always in the world and only passes from one part of matter to another, agreeably to the laws of nature and the beautiful pre-established order" (*ibid*.: 'Leibniz's First Paper' 1.4, pp. 675–676).

[21] Leibniz was clear on this. Neither substance nor accident (i.e., particularized property) transfers, literally speaking. There is literally no *transfer* of either. What there is is a detailed and deep and thoroughgoing pre-established harmony of internal changes in substances (*ibid., passim*).

[22] *Cf*. Leibniz's claim: "We must recognize that force is something absolutely real even in created substances, but that space, time, and motion have something akin to a mental construction [*de ente rationis*]" (*ibid., Specimen Dynamicum, op. cit*., p. 445).

[23] This issue was raised for me by my associate Earl Conee, with whom I discussed the ideas of this essay before I wrote it down.

[24] Leibniz was very emphatic on the simplicity and comprehensiveness of the laws of nature. He based that view on the idea that God's perfection could not allow him to create other types of laws. Naturally, he was very firm on what in modern terminology can be described as the conservation of energy. He put it differently, his favorite formula being something like "the whole effect is always equal to the full cause" (*ibid*., p. 444).

After I had fully conceived this essay, but before I finished writing it, I read Henry Margenau's *The Nature of Physical Reality*, McGraw-Hill, New York, 1950, Chap. 19, 'Causality'. Margenau offers as his most central claim the following: "*Causality holds if the laws of nature (differential equations) governing closed systems do not contain the time variable in explicit form*" (p. 405; original emphasis). He later on speaks of his analysis of cause. Hence, we must understand his "if" in the preceding quotation as an analyzing 'if and only if'. He also regards causality so analyzed as a necessary condition for the conservation principle, but not as equivalent to it (p. 411). Margenau's discussion is very illuminating: it comes up from the side of physics. Mine comes from the ontological side. Undoubtedly, I must have read that chapter some time in 1952, when I bought my copy of the book. I hypothesize that it has unconsciously guided my reflections on causality.

The general principle of conceptual and computational simplicity behind the principles of causity conservation must be distinguished from more specific and substantive principles of simplicity. The general principle underlies these specific principles too. One most important example is Hamilton's principle: "Of all possible paths along which a dynamical system may move from one point to another, within a specified time interval and consistent with any constraints, the actual path followed is that which minimizes the integral with respect to time of the difference between the kinetic and potential energies".

[25] I am most grateful to Walter Bethel for having presented me with eleven single-spaced pages of corrections of typos, grammar, style, coherence, and of deep criticisms. Unfortunately I could not, because of the length of the paper, avail myself of all his suggestions.

Note Added in the Proof. A first, fascinating study comparing Leibniz's and Hume's views on causality is Hidé Ishiguro, *Pre-established Harmony versus Constant Conjunction: A Reconsideration of the Distinction Between Rationalism and Empiricism* (British Academy, Oxford University Press, 1977).

SYDNEY SHOEMAKER

CAUSALITY AND PROPERTIES

I

It is events, rather then objects or properties, that are usually taken by philosophers to be the terms of the causal relationship. But an event typically consists of a change in the properties or relationships of one or more objects, the latter being what Jaegwon Kim has called the "constituent objects" of the event.[1] And when one event causes another, this will be in part because of the properties possessed by their constituent objects. Suppose, for example, that a man takes a pill and, as a result, breaks out into a rash. Here the cause and effect are, respectively, the taking of the pill and the breaking out into a rash. Why did the first event cause the second? Well, the pill was penicillin, and the man was allergic to penicillin. No doubt one could want to know more – for example, about the biochemistry of allergies in general and this one in particular. But there is a good sense in which what has been said already explains why the one event caused the other. Here the pill and the man are the constituent objects of the cause event, and the man is the constituent object of the effect event. Following Kim we can also speak of events as having "constituent properties" and "constituent times". In this case the constituent property of the cause event is the relation expressed by the verb 'takes', while the constituent property of the effect event is expressed by the predicate 'breaks out into a rash'. The constituent times of the events are their times of occurrence. Specifying the constituent objects and properties of the cause and effect will tell us what these events consisted in, and together with a specification of their constituent times will serve to identify them; but it will not, typically, explain why the one brought about the other. We explain this by mentioning certain properties of their constituent objects. Given that the pill was penicillin, and that the man was allergic to penicillin, the taking of the pill by the man was certain, or at any rate very likely, to result in an allergic response like a rash. To take another example, suppose a branch is blown against a window and breaks it. Here the constituent objects include the branch and the window, and the causal relationship holds because of, among other things, the massiveness of the one and the fragility of the other.

It would appear from this that any account of causality as a relation

Peter van Inwagen (ed.), Time and Cause, 109–135.
Copyright © 1980 *by D. Reidel Publishing Company.*

between events should involve, in a central way, reference to the properties of the constituent objects of the events. But this should not encourage us to suppose that the notion of causality is to be analyzed away, in Humean fashion, in terms of some relationship between properties – for example, in terms of regularities in their instantiation. For as I shall try to show, the relevant notion of a property is itself to be explained in terms of the notion of causality in a way that has some strikingly non-Humean consequences.

II

Philosophers sometimes use the term 'property' in such a way that for every predicate *F* true of a thing there is a property of the thing which is designated by the corresponding expression of the form 'being *F*'. If 'property' is used in this broad way, every object will have innumerable properties that are unlikely to be mentioned in any causal explanation involving an event of which the object is a constituent. For example, my typewriter has the property of being over one hundred miles from the current heavyweight boxing champion of the world. It is not easy to think of a way in which its having this property could help to explain why an event involving it has a certain effect, and it seems artificial, at best, to speak of my typewriter's acquisition of this property as one of the causal effects of the movements of the heavyweight champion.

It is natural, however, to feel that such properties are not 'real' or 'genuine' properties. Our intuitions as to what are, and what are not, genuine properties are closely related to our intuitions as to what are, and what are not, genuine changes. A property is genuine if and only if its acquisition or loss by a thing constitutes a genuine change in that thing. One criterion for a thing's having changed is what Peter Geach calls the "Cambridge criterion". He formulates this as follows: "The thing called 'x' has changed if we have '$F(x)$ at time t' true and '$F(x)$ at time t^1' false, for some interpretations of 'F', 't', and 't^1' ".[2] But as Geach points out, this gives the result that Socrates undergoes a change when he comes to be shorter than Theaetetus in virtue of the latter's growth, and even that he undergoes a change every time a fresh schoolboy comes to admire him. Such 'changes', those that intuitively are not genuine changes, Geach calls "mere 'Cambridge' changes". For Geach, real changes are Cambridge changes, since they satisfy the Cambridge criterion, but some Cambridge changes, namely those that are *mere* Cambridge changes, fail to be real changes. Since it is mere Cambridge changes, rather than Cambridge changes in general, that are to be contrasted with real or genuine changes, I shall introduce the hyphenated expression 'mere-Cambridge' to characterize these. And I shall apply the terms 'Cambridge' and 'mere-Cambridge' to properties as well as to changes. Mere-Cambridge properties will include such properties as being 'grue' (in Nelson Goodman's sense), historical properties like being over twenty years old and having been slept in by George Washington, relational properties like being fifty miles south of burning barn,[3] and such properties as being such that Jimmy Carter is President of the United States.

It is worth mentioning that in addition to distinguishing between real and mere-Cambridge properties and changes, we must also distinguish between real and mere-Cambridge resemblance or similarity, and between real and mere-Cambridge differences. Cambridge similarities hold in virtue of the sharing of Cambridge properties. And mere-Cambridge similarities hold in virtue of the sharing of mere-Cambridge properties: there is such a similarity between all grue things; there is one between all things fifty miles south of a burning barn; there is one between all beds slept in by George Washington; and there is one between all things such that Jimmy Carter is President of the United States. It will be recalled that the notion of similarity, or resemblance, plays a prominent role in Hume's account of causality. His first definition of *cause* in the *Treatise* is "an object precedent and contiguous to another, and where all the objects resembling the former are plac'd in a like relation of priority and contiguity to those objects, that resemble the latter."[4] Hume clearly regarded the notion of resemblance as quite unproblematical and in no need of elucidation.[5] Yet it is plain that he needs a narrower notion of resemblance than that of Cambridge resemblance if his definition of causality is to have the desired content. Cambridge resemblances are too easily come by; any two objects share infinitely many Cambridge properties, and so 'resemble' one another in infinitely many ways. There are also infinitely many Cambridge differences between any two objects. What Hume needs is a notion of resemblance and difference which is such that some things resemble a given thing more than others do, and such that some things may resemble a thing exactly (without being numerically identical to it) while others resemble it hardly at all. Only 'real' or 'genuine' resemblance will serve his purposes. If it turns out, as I think it does, that in order to give a satisfactory account of the distinction between real and mere-Cambridge properties, changes, similarities, and difference we must make use of the notion of causality, the Humean project of defining causality in terms of regularity or 'constant conjunction', notions that plainly involve the notion of resemblance, is seriously undermined.

I have no wish to legislate concerning the correct use of the terms 'property', 'change', 'similar', and so forth. It would be rash to claim that the accepted use of the term 'property' is such that what I have classified as mere-Cambridge properties are not properties. But I do think that we have *a* notion of what it is to be a property which is such that this is so – in other words, which is such that not every phrase of the form 'being so and so' stands for a property which something has just in case the corresponding predicate of the form 'is so and so' is true of it, and is such that sometimes a predicate is true of a thing, not because (or only because) of any properties *it* has, but because something else, perhaps something related to it in certain ways, has certain properties. It is this narrow conception of what it is to be a property, and the correlative notions of change and similarity, that I am concerned to elucidate in this paper. (I should mention that I am concerned here only with the sorts of properties with respect to which change is possible; my account is not intended to apply to such properties of numbers as being even and being prime.)

III

John locke held that "powers make a great part of our complex ideas of substances."[6] And there is one passage in which Locke seems to suggest that all qualities of substances are powers; he says, in explanation of his usage of the term 'quality', that "the power to produce any idea in our mind, I call *quality* of the subject wherein that power is ".[7] This suggests a theory of properties, namely that properties are causal powers, which is akin to the theory I shall be defending. As it happens, this is not Locke's view. If one ascribed it to him on the basis of the passage just quoted, one would have to ascribe to him the view that all qualities are what he called 'secondary qualities' – powers to produce certain mental effects ('ideas') in us. But Locke recognized the existence of powers that are not secondary qualities, namely powers (for example, the power in the sun to melt wax) to produce effects in material objects. These have been called 'tertiary qualities'. And he distinguished both of these sorts of powers from the 'primary qualities' on which they 'depend'. Nevertheless, the view which Locke's words unintentionally suggest is worth considering.

What would seem to be the same view is sometimes put by saying that all properties are dispositional properties. But as thus formulated, this view seems plainly mistaken. Surely we make a distinction between dispositional and nondispositional properties, and can mention paradigms of both sorts. Moreover, it seems plain that what dispositional properties something has, what powers it has, depends on what nondispositional properties it has – just as Locke thought that the powers of things depend on their primary qualities and those of their parts.

In fact, I believe, there are two different distinctions to be made here, and these are often conflated. One is not a distinction between kinds of *properties* at all, but rather a distinction between kinds of *predicates*. Sometimes it belongs to the meaning, or sense, of a predicate that if it is true of a thing then under certain circumstances the thing will undergo certain changes or will produce certain changes in other things. This is true of what are standardly counted as dispositional predicates, for example, 'flexible', 'soluble', 'malleable', 'magnetized', and 'poisonous'. Plainly not all predicates are of this sort. Whether color predicates are is a matter of controversy. But whatever we say about this, it seems plain that predicates like 'square', 'round' and 'made of copper' are not dispositional in this sense. There are causal powers associated with being made of copper – for example, being an electrical conductor. But presumably this association is not incorporated into the meaning of the term 'copper'.

The first distinction, then, is between different sorts of predicates, and I think that the term 'dispositional' is best employed as a predicate of predicates, not of properties. A different distinction is between powers, in a sense I am about to explain, and the properties in virtue of which things have the powers they have.[8] For something to have a power, in this sense, is for it to be such that its presence in circumstances of a particular sort will have certain effects.[9] One can think of such a power as a function from circumstances to effects. Thus if something is poisonous its presence in someone's body will produce death or illness; in virtue of this, being poisonous is a power. Here it is possible for things to have the same power in virtue of having very different properties. Suppose that one poisonous substance kills by affecting the heart, while another kills by directly affecting the nervous system and brain. They produce these different effects in virtue of having very different chemical compositions. They will of course differ in their powers as well as in their properties, for one will have the power to produce certain physiological effects in the nervous system, while the other will have the power to produce quite different physiological effects in the heart. But there is one power they will share, in virtue of having these different powers, namely that of producing death if ingested by a human being. Properties here play the role, vis-à-vis powers, that primary qualities play in Locke; it is in virtue of a thing's properies that the thing has the powers (Locke's secondary and tertiary qualities) that it has.

There is a rough correspondence between this distinction between powers and properties and the earlier distinction between dispositional and nondispositional predicates. By and large, dispositional predicates ascribe powers while nondispositional monadic predicates ascribe properties that are not powers in the same sense.

IV

On the view of properties I want to propose, while properties are typically not powers of the sort ascribed by dispositional predicates, they are related to such powers in much the way that such powers are related to the causal effects which they are powers to produce. Just as powers can be thought of as functions from circumstances to causal effects, so the properties on which powers depend can be thought of as functions from properties to powers (or, better, as functions from sets of properties to sets of powers). One might even say that properties are second-order powers; they are powers to produce first-order powers (powers to produce certain sorts of events) if combined with certain other properties. But the formulation I shall mainly employ is this: what makes a property the property it is, what determines its identity, is its potential for contributing to the causal powers of the things that have it. This means, among other things, that if under all possible circumstances properties *X* and *Y* make the same contribution to the causal powers of the things that have them, *X* and *Y* are the same property.

To illustrate this, let us take as our example of a property the property of being 'knife-shaped' – I shall take this to be a highly determinate property which belongs to a certain knife in my kitchen and to anything else of exactly the same shape. Now if all that I know about a thing is that it has this property, I know nothing about what will result from its presence in any circumstances. What has the property of being knife-shaped could be a knife, made of steel, but it could instead be a piece of balsa wood, a piece of butter, or even an oddly-shapped cloud of some invisible gas. There is no power which necessarily belongs to all and only the things having this property. But if this property is combined with the property of being knife-sized and the property of being made of steel, the object having these properties will necessarily have a number of powers. It will have the power of cutting butter, cheese and wood, if applied to these substances with suitable pressure, and also the power of producing various sorts of sense impressions in human beings under appropriate observational conditions, and also the power of leaving an impression of a certain shape if applied to soft wax and then withdrawn, and so on. The combination of the property of being knife-shaped with the property of being made of glass will result in a somewhat different set of powers, which will overlap with the set which results from its combination with the property of being made of steel. Likewise with its combination with the property of being made of wood, the property of being made of butter, and so on.

Let us say that an object has power *P* conditionally upon the possession of the properties in set *Q* if it has some property *r* such that having the properties in *Q* together with *r* is causally sufficient for having *P*, while having the properties in *Q* is not by itself causally sufficient for having *P*. Thus, for example, a knife-shaped object has the power of cutting wood conditionally upon being knife-sized and made of steel; for it is true of knife-shaped things, but not of things in general, that if they are knife-sized and made of steel they will have the power to cut wood. When a thing has a power conditionally upon the possession of certain properties, let us say that this amounts to its having a *conditional power*. Our knife-shaped object has the conditional power of being able to cut wood if knife-sized and made of steel. The identity condition for conditional powers is as follows: if *A* is the conditional power of having power *P* conditionally upon having the properties in set *Q*, and *B* is the conditional power of having P' conditionally upon having the properties in set Q', then *A* is identical to *B* just in case *P* is identical to P' and *Q* is identical to Q'. Having introduced this notion of a conditional power, we can express my view by saying that properties are clusters of conditional powers. (I shall count powers *simpliciter* as a special case of conditional powers.) I have said that the identity of property is determined by its causal potentialities, the contributions it is capable of making to the causal powers of things that have it. And the causal potentialities that are essential to a property correspond to the conditional powers that make up the cluster with which the property can be identified; for a property to have a causal potentiality is for it to be such that whatever has it has a certain conditional power.

This account is intended to capture what is correct in the view that properties just are powers, or that all properties are dispositional, while acknowledging the truth of a standard objection to that view, namely that a thing's powers or dispositions are distinct from, because 'grounded in', its intrinsic properties.[10]

Before I give my reasons for holding this view, I should mention one prima facie objection to it. Presumably the property of being triangular and the property of being trilateral do not differ in the contributions they make to the causal powers of the things that have them, yet it is natural to say that these, although necessarily coextensive, are different properties. It seems to me, however, that what we have good reason for regarding as distinct are not these properties, as such, but rather the concepts of triangularity and trilaterality, and the meanings of the expressions 'triangular' and 'trilateral'. If we abandon, as I think we should, the idea that properties are the meanings of predicate expressions, and if we are careful to distinguish concepts from what they are concepts of, I see no insuperable obstacle to regarding the properties themselves as identical.

V

My reasons for holding this theory of properties are, broadly speaking, epistemological. Only if some causal theory of properties is true, I believe, can it be explained how properties are capable of engaging our knowledge, and our language, in the way they do.

We know and recognize properties by their effects, or, more precisely, by the effects of the events which are the activations of the causal powers which things have in virtue of having the properties. This happens in a variety of ways. Observing something is being causally influenced by it in certain ways. If the causal potentialities involved in the possession of property are such that there is a fairly direct causal connection between the possession of it by an object and the sensory states of an observer related to that object in certain ways, e. g., looking at it in good light, we say that the property itself is observable. If the relationship is less direct, e. g., if the property can affect the sensory states of the observer only by affecting the properties of something else which the observer observes, a scientific instrument, say, we speak of inferring that the thing has the property from what we take to be the effects of its possession. In other cases we conclude that something has a property because we know that it has other properties which we know from other cases to be correlated with the one in question. But the latter way of knowing about the properties of things is parasitic on the earlier ways; for unless the instantiation of the property had, under some circumstances, effects from which its existence could be concluded, we could never discover laws or correlations that would enable us to infer its existence frrom things other than its effects.

Suppose that the identity of properties consisted of something logically independent of their causal potentialities. Then it ought to be possible for there to be properties that have no potential whatever for contributing to causal powers, i. e., are such that under no conceivable circumstances will their possession by a thing make any difference to the way the presence of that thing affects other things or to the way other things affect it. Further, it ought to be possible that there be two or more different properties that make, under all possible circumstances, exactly the same contribution to the causal powers of the things that have them. Further, it ought to be possible that the potential of a property for contributing to the production of causal powers might change over time, so that, for example, the potential possessed by property *A* at one time is the same as that possessed by property *B* at a later time, and that possessed by property *B* at the earlier time is the same as that possessed by property *A* at the latter time. Thus a thing might undergo radical change with respect to its properties without undergoing any change in its causal powers, and a thing might undergo radical change in its causal powers without undergoing any change in the properties that underlie these powers.

The supposition that these possibilities are genuine implies, not merely (what might seem harmless) that various things might be the case without its being in any way possible for us to know that they are, but also that it is impossible for us to know various things which we take ourselves to know. If there can be properties that have no potential for contributing to the causal powers of the things that have them, then nothing could be good evidence that the overall resemblance between two things is greater than the overall resemblance between two other things; for even if *A* and *B* have closely resembling effects on our senses and our instruments while *C* and *D* do not, it might be (for all we know) that *C* and *D* share vastly more properties of the causally impotent kind than do *A* and *B*. Worse, if two properties can have exactly the same potential for contributing to causal powers, then it is impossible for us even to know (or have any reason for believing) that two things resemble one another by sharing a single property. Moreover, if the properties and causal potentialities of a thing can vary independently of one another, then it is impossible for us to know (or have any good reason for believing) that something has retained a property over time, or that something has undergone a change with respect to the properties that underlie its causal powers. On these suppositions, there would be no way in which a particular property could be picked out so as to have a name attached to it; and even if, *per impossibile*, a name did get attached to a property, it would be impossible for anyone to have any justification for applying the name on particular occasions.

It may be doubted whether the view under attack has these disastrous epistemological consequences. Surely, it may be said, one can hold that it is a contingent matter that particular properties have the causal potentialities they have, and nevertheless hold, compatibly with this, that there are good theoretical reasons for thinking that as a matter of fact different properties differ in their causal potentialities, and that any given property retains the same potentialities over time. For while it is logically possible that the latter should not be so, according to the contingency view, the simplest hypothesis is that it is so; and it is reasonable to accept the simplest hypothesis compatible with the data.

Whatever may be true in general of appeals to theoretical simplicity, this one seems to me extremely questionable. For here we are not really dealing with an explanatory hypothesis at all. If the identity of properties is made independent of their causal potentialities, then in what sense do we explain sameness or difference of causal potentialities by positing sameness or difference of properties? There are of course cases in which we explain a constancy in something by positing certain underlying constancies in its properties. It is genuinely explanatory to say that something retained the same causal power over time because certain of its properties remained the same. And this provides, *ceteris paribus*, a simpler, or at any rate more plausible, explanation of the constancy than one that says that the thing first had one set of underlying properties and then a different set, and that both sets were sufficient to give it that particular power. For example, if the water supply was poisonous all day long, it is more plausible to suppose that this was due to the presence in it of one poisonous substance all day rather than due to its containing cyanide from morning till noon and strychnine from noon till night. But in such cases we presuppose that the underlying property constancies carry with them constancies in causal potentialities, and it is only on this presupposition that positing the underlying constancies provides the simplest explanation of the constancy to be explained. Plainly this presupposition cannot be operative if what the 'inference to the best explanation' purports to explain is, precisely, that sameness of property goes with sameness of causal potentialities. It is not as if a property had the causal potentialities in question as a result of having yet *other* causal potentialities, the constancy of the latter explaining the constancy of the former. This disassociation of property identity from identity of causal potentiality is really an invitation to eliminate reference to properties from our explanatory hypotheses altogether; if it were correct then we could, to use Wittgenstein's metaphor, 'divide through' by the properties and leave the explanatory power of what we say about things untouched.

It might be objected that even if my arguments establish that the causal potentialities of a genuine property cannot change over time, they do not establish that these causal potentialities are essential to that property, in the sense of belonging to it in all possible worlds. The immutability of properties with respect to their causal potentialities, it might be said, is simply a consequence of the immutability of laws – of the fact that it makes no sense to speak of a genuine law holding at one time and not at another. And from the fact that the laws governing a property cannot change over time it does not follow, it may be said, that the property cannot be governed by different laws in different possible worlds.

Let me observe first of all that in conceding that the immutability of the causal potentialities of genuine properties is a consequence of the immutability of laws, the objection concedes a large part of what I want to maintain. It is not true in general of mere-Cambridge properties that their causal potentialities cannot change over time; for example, this is not true of *grueness* on the Barker-Achinstein definition of *grue,* where something is grue just in case it is green and the time is before *T* (say 2000 A.D.) or it is blue and the time is *T* or afterwards.[11] That genuine properties are marked off from mere-Cambridge properties by their relation to causal laws (and that it is nonsense to speak of a world in which it is the mere-Cambridge properties rather than the genuine ones that are law-governed in way that makes their causal potentialities immutable) is a central part of my view.

There is, moreover, a prima facie case for saying that the immutability of the causal potentialities of a property does imply their essentiality; or in other words, that if they cannot vary across time, they also cannot vary across possible worlds. Most of us do suppose that *particulars* can (or do) have different properties in different possible worlds. We suppose, for example, that in some possible worlds I am a plumber rather than a philosopher, and that in some possible worlds my house is painted yellow rather than white. But it goes with this that particulars can change their properties over time. It is possible that I, the very person who is writing this paper, might have been a plumber, because there is a possible history in which I start with the properties (in this case relational as well as intrinsic) which I had at some time in my actual history, and undergo a series of changes which result in my eventually being a plumber. If I and the world were never such that it was then possible for me to *become* a plumber, it would not be true that I might have been a plumber, or (in other words) that there is a possible world in which I am one. There is, in short, a close linkage between identity across time and identity across possible worlds; the ways in which a given thing can be different in different possible worlds depend on the ways in which such a thing can be different at different times in the actual world. But now let us move from the case of particulars to that of properties. There is no such thing as tracing a property through a series of changes in its causal potentialities – not if it is a genuine property, i. e., one of the sort that figures in causal laws. And so there is no such thing as a possible history in which a property starts with the set of causal potentialities it has in the actual world and ends with a different set. To say the least, this calls into question the intelligibility of the suggestion that the very properties we designate with words like 'green', 'square', 'hard', and so on, might have had different causal potentialities than they in fact have.

However, this last argument is not conclusive. My earlier arguments, if sound, establish that there is an intimate connection between the identity of a property and its causal potentialities. But it has not yet been decisively established that *all* of the causal potentialities of a property are essential to it. The disastrous epistemological consequences of the contingency view would be avoided if for each property we could identify a proper subset of its causal potentialities that are essential to it and constitutive of it, and this would permit some of a property's causal potentialities, those outside the essential cluster, to belong to it contingenty, and so not belong to it in some other possible worlds. There would, in this case, be an important difference between the trans-world indentity of properties and that of particulars – and it is a difference which there is in my own view as well. If, as I believe, the assertion that a certain particular might have had different properties than it does in the actual world (that in some other possible world it does have those properties) implies that there is a possible history 'branching off' from the history of the actual world in which it acquires those properties, this is because there is, putting aside historical properties and 'identity properties' (like being identical to Jimmy Carter), no subset of the properties of such a thing which constitutes an individual essence of it, i. e., is such that, in any possible world, having the properties in that subset is necessary *and sufficient* for being that particular thing. To put this otherwise, the reason why the possible history in which the thing has different properties must be a branching-off from the history of the actual world is that the individual essence of a particular thing must include historical properties. Now I am not in a position to object to the suggestion that properties differ from particulars in having individual essences which do not include historical properties and which are sufficient for their identification across possible worlds; for I hold that the totally of a property's causal potentialities constitute such an individual essence. So a possible alternative to my view is one which holds that for each property there is a proper subset of its causal potentialities that constitute its individual essence. Such a view has its attractions, and is compatible with much of what I say in this paper; in particular, it is compatible with the claim that within any possible world properties are identical just in case they have the same causal potentialities. But I shall argue in Section IX that this view is unworkable, and that there is no acceptable alternative to the view that all of the causal potentialities of a property are essential to it.

VI

As was intended, my account of properties does not apply to what I have called mere-Cambridge properties. When my table acquired the property of being such that Gerald Ford is President of the United States, which it did at the time Nixon resigned from the presidency, this presumably had no effect on its causal powers. Beds that were slept in by George Washington may command a higher price than those that lack this historical property, but presumably this is a result, not of any causal potentialities in the beds themselves, but of the historical beliefs and interests of those who buy and sell them. And grueness, as defined by Goodman, is not associated in the way greenness and blueness are with causal potentialities. (In this sense, which differs from that invoked in Section V, something is grue at a time just in case it is green at that time and is first examined before *T*, say, 2000 A.D., or is blue at that time and is not first examined before *T*.) It can happen that the only difference between something that is grue and something that is not is that one of them has and the other lacks the historical property of being (or having been) first examined before the time *T* mentioned in Goodman's definition of *grue*; and presumably this does not in itself make for any difference in causal potentialities. It can also happen that two things share the property of being grue in virtue of having properties that have different potentialities – that is, in virtue of one of them being green (and examined before *T*) and the other being blue (and not so examined).

There is an epistemological way of distinguishing genuine and mere-Cambridge properties that is prima facie plausible. If I wish to determine whether an emerald is green at *t*, the thing to do, if I can manage it, is to examine the emerald at *t*. But examination of a table will not tell me it is such that Gerald Ford is President of the United States, or whether it is fifty miles south of a burning barn. And if I am ignorant of the date, or if *t* is after *T* (the date in Goodman's definition), examination of an emerald will not tell me whether it is grue. Likewise, while scrutiny of a bed may reveal a plaque claiming that it was slept in by George Washington, it will not tell me whether this claim is true. Roughly, if a question about whether thing has a property at a place and time concerns a genuine nonrelational property, the question is most directly settled by observations and tests in the vicinity of that place and time, while if it concerns a mere-Cambridge property it may be most directly settled by observations and tests remote from that place and time, and observations and tests made at that place and time will either be irrelevant (as in the case of the property of being such that Jimmy Carter is president) or insufficient to settle the question (as in the case of grue).

It would be difficult to make this into a precise and adequate criterion of genuineness of property, and I do not know whether this could be done. But I think that to the extent that it is adequate, its adequacy is explained by my account of properties in terms of causal powers. Properties reveal their presence in actualizations of their causal potentialities, a special case of this being the perception of a property. And the most immediate and revealing effects of an object's having a property at a particular place and time are effects that occur in the immediate vicinity of that place and time. To be sure, we cannot rule out on purely philosophical grounds the possibility of action at a spatial and/or temporal distance. And the more prevalent such action is, the less adequate the proposed epistemological criterion will be. But there do seem to be conceptual limitations on the extent to which causal action can be at a spatial or temporal distance. It is doubtful, to say the least, whether there could be something whose causal powers are *all* such that whenever any of them is activated the effects of its activation are spatially remote from the location of the thing at that time, or occur at times remote from the time of activation.

Causation and causal powers are as much involved in the verification of ascriptions of mere-Cambridge properties as in the verification of ascriptions of genuine ones. But in the case of mere-Cambridge properties some of the operative causal powers will either belong to something other than the object to which the property is ascribed, or will belong to that object at a time other than that at which it has that property. Thus if I verify that a man has the property of being fifty miles south of a burning barn, it will be primarily causal powers of the barn, and of intervening stretch of land (which, we will suppose, I measure), rather than the causal powers of the man, that will be responsible for my verifying observations.

VII

It will not have escaped notice that the account of properties and property identity I have offered makes free use of the notion of a property and the notion of property identity. It says, in brief, that properties are identical, whether in the same possible world or in different ones, just in case their coinstantiation with the same properties gives rise to the same powers. This is, if anything, even more circular than it looks. For it crucially involves the notion of sameness of powers, and this will have to be explained in terms of sameness of circumstances and sameness of effects, the notions of which both involve the notion of sameness of property. And of course there was essential use of the notion of a property in my explanation of the notion of a conditional power.

It is worth observing that there is a distinction between kinds of powers that corresponds to the distinction, mentioned earlier, between genuine and mere-Cambridge properties.[12] Robert Boyle's famous example of the key can be used to illustrate this.[13] A particular key on my key chain has the power of opening locks of a certain design. It also has the power of opening my front door. It could lose the former power only by undergoing what we would regard as real change, for example, a change in its shape. But it could lose the latter without undergoing such a change; it could so do in virtue of the lock on my door being replaced by one of a different design. Let us say that the former is an intrinsic power and the latter a mere-Cambridge power. It is clear that in my account of properties the word 'power' must refer only to intrinsic powers. For if it refers to mere-Cambridge powers as well, then what seems clearly to be a mere-Cambridge property of my key, namely being such that my door has a lock of a certain design, will make a determinate contribution to its having the powers it has, and so will count as a genuine property of it. But it seems unlikely that we could explain the distinction between intrinsic and mere-Cambridge powers without making use of the notion of a genuine change and that of a genuine property. And so again my account of the notion of a property in terms of the notion of a power can be seen to be circular.

How much do these circularities matter? Since they are, I think, unavoidable, they preclude a reductive analysis of the notion of a property in terms of the notion of causality. But they by no means render my account empty. The claim that the causal potentialities of a property are essential to it, and that properties having the same causal potentialities are identical, is certainly not made vacuous by the fact that the explanation of the notion of a causal potentiality, or a conditional power, must invoke the notion of a property. As I see it, the notion of a property and the notion of a causal power belong to a system of internally related concepts, no one of which can be explicated without the use of the others. Other members of the system are the concept of an event, the concept of similarity, and the concept of a persisting substance. It can be worthwhile, as a philosophical exercise, to see how far we can go in an attempt to reduce one of these concepts to others – for both the extent of our success and the nature of our failures can be revealing about the nature of the connections between the concepts. But ultimately such attempts must fail. The goal of philosophical analysis, in dealing with such concepts, should not be reductive analysis but rather the charting of internal relationships. And it is perfectly possible for a 'circular' analysis to illuminate a network of internal relationships and have philosophically interesting consequences.

VIII

According to the theory of properties I am proposing, all of the causal potentialities possessed by a property at any time in the actual world are essential to it and so belong to it at all times and in all possible worlds. This has a very strong consequence, namely that causal necessity is just a species of logical necessity. If the introduction into certain circumstances of a thing having certain properties causally necessitates the occurrence of certain effects, then it is impossible, logically impossible, that such an introduction could fail to have such an effect, and so logically necessary that it has it. To the extent that causal laws can be viewed as propositions describing the causal potentialities of properties, it is impossible that the same properties should be governed by different causal laws in different possible worlds, for such propositions will be necessarily true when true at all.

It is not part of this theory, however, that causal laws are analytic or knowable a priori. I suppose that it is analytic that flexible things bend under suitable pressure, that poisonous things cause injury to those for whom they are poisonous, and so on. But I do not think that it is analytic that copper is an electrical conductor, or that knife-shaped things, if knife-sized and made of steel, are capable of cutting butter. Nor does it follow from the claim that such truths are necessary that they are analytic. Kripke has made a compelling case for the view that there are propositions that are necessary a posteriori, that is, true in all possible worlds but such that they can only be known empirically.[14] And such, according to my theory, is the status of most propositions describing the causal potentialities of properties. The theory can allow that our knowledge of these potentialities is empirical, and that it is bound to be only partial. But in order to show how, in the theory, such empirical knowledge is possible, I must now bring out an additional way in which the notion of causality is involved in the notion of a property.

One of the formulations of my theory says that every property is a cluster of conditional powers. But the converse does not seem to me to hold; not every cluster of conditional powers is a property. If something is both knife-shaped and made of wax, then it will have, among others, the following conditional powers: the power of being able to cut wood conditionally upon being knife-sized and made of steel (this it has in virtue of being knife-shaped), and the power of being malleable conditionally upon being at a temperature of 100°F (this it has in virtue of being made of wax). Intuitively, these are not common components of any single property. By contrast, the various conditional powers a thing has in virtue of being knife-shaped – for example, the power of being able to cut wood conditionally upon being knife-sized and made of steel, the power of being able to cut butter conditionally upon being knife-sized and made of wood, the power of having a certain visual appearance conditionally upon being green, the power of having a certain other visual appearance conditionally upon being red, and so on – are all constituents of a single property, namely the property of being knife-shaped. The difference, I think, is that in the one case the set of conditional powers has, while in the other it lacks, a certain kind of causal unity. I shall now try to spell out the nature of this unity.

Some subsets of the conditional powers which make up a genuine property will be such that it is a consequence of causal laws that whatever has any member of the subset necessarily has all of its members. Thus, for example, something has the power of leaving a six-inch-long knife-shaped impression in soft wax conditionally upon being six inches long if and only if it has the power of leaving an eight-inch-long knife-shaped impression in soft wax conditionally upon being eight inches long. Now some conditional powers will belong to more than one property cluster; thus, for example, there are many different shape properties that give something the power of being able to cut wood conditionally upon being made of steel. But where a conditional power can be shared by different properties in this way, it will belong to a particular property cluster only if there is another member of that cluster which is such that it is a consequence of causal laws that whatever has that other member has the conditional power in question. And at the core of each cluster there will be one or more conditional powers which are such that as a consequence of causal laws whatever has any of them has all of the conditional powers in the cluster. For example, if something has, conditionally upon being made of steel, the power of leaving a knife-shaped impression in soft wax, then it cannot fail to be knife-shaped, and so cannot fail to have all of the other conditional powers involved in being knife-shaped. I suggest, then, that conditional powers *X* and *Y* belong to the same property if and only if it is a consequence of causal laws that either (1) whatever has either of them has the other, or (2) there is some third conditional power such that whatever has it has both *X* and *Y*.

Returning now to the conditional power of being able to cut wood conditionally upon being made of steel and the conditional power of being malleable conditionally upon being at a temperature of 100°F, it seems to me that these do not qualify under the proposed criterion as belonging to a common property. It is obviously not true that whatever has one of them must have the other. And it does not appear that there is any third conditional power which is such that whatever has it must have the two conditional powers in question.[15]

If I am right in thinking that the conditional powers constituting a property must be causally unified in the way indicated, it is not difficult to see how knowledge of the causal potentialities of properties can develop empirically. The behavior of objects, that is, the displays of their powers, will reveal that they have certain conditional powers. Once it is discovered that certain conditional powers are connected in a lawlike way, we can use these to "fix the reference" of a property term to the cluster containing those conditional powers and whatever other conditional powers are related to them in the appropriate lawlike relationships.[16] And we can then set about to determine empirically what the other conditional powers in the cluster are.

IX

As I observed earlier, my theory appears to have the consequence that causal laws are logically necessary, and that causal necessity is just a species of logical necessity. While to some this may be an attractive consequence, to many it will seem counterintuitive. It does seem to most of us that we can conceive of possible worlds which resemble the actual world in the kinds of properties that are instantiated in them, but differ from it in the causal laws that obtain. My theory must maintain either that we cannot really conceive of this or that conceivability is not proof of logical possibility.

Anyone who finds both of these alternatives unacceptable, but is persuaded by the arguments in Section V that the identity of properties is determined by their causal potentialities, will look for ways of reconciling that conclusion with the view that there can be worlds in which some of the causal laws are different from, and incompatible with, those that obtain in the actual world. I want now to consider two ways in which one might attempt to achieve such a reconciliation. First, it might be held that while propositions describing the causal potentialities of properties are necessarily true if true at all, there are other lawlike propositions, namely those asserting lawlike connections between conditional powers, which are contingent and so true in some possible worlds and false in others. According to this view, when we seem to be conceiving of worlds in which the same properties are governed by different laws, what we are really conceiving of are worlds in which the same conditional powers stand to one another in different lawlike connections than they do in the actual world, and so are differently clustered into properties. Second, it might he held that my condition for the identity of properties across possible worlds is too strict. The theory I have advanced might be called the 'total cluster theory'; it identifies a property with a cluster containing all of the conditional powers which anything has in virtue of having that property, and maintains that in any possible world anything that has that property must have all of the members of that cluster. One might attempt to replace this with a 'core cluster theory', which identifies the property with some proper subset of the conditional powers something has in virtue of having that property. On this theory, it is only some of the causal potentialities possessed by a property in the actual world, namely those constituted by the conditional powers in its core cluster, that are essential to it – so it is possible for the same property to have somewhat different causal potentialities in different possible worlds, because of different laws relating the conditional powers in its core cluster with other conditional powers.

I do not believe, however, that either of these attempted reconciliations is sucessful. The first involves the suggestion that it is at least sometimes a contingent matter whether two conditional powers belong to the same property, and hence that there could be a world in which some of the same conditional powers are instantiated as in this world, but in which, owing to the holding of different laws, these are differently clustered into properties. The difficulty with this is that the specification of a conditional power always involves, in two different ways, reference to properties that are instantiated in our world and which, *ex hypothesi*, would not be instantiated in the alternative world in question. It involves reference to the properties on which the power is conditional, and also to the properties in the instantiation of which the exercise of the power would result. For example, one of the conditional powers in the property of being knife-shaped is the power, conditionally upon being made of steel, of leaving a knife-shaped impression if pressed into soft wax and then withdrawn. This conditional power, although not by itself identical to the property of being knife-shaped, could not be exercised without that property being instantiated. Neither could it be exercised without the property of being made of steel being instantiated. And a conditional power could not be instantiated in a world in which the causal laws would not allow an exercise of it. So in general, a conditional power could not be instantiated in a world in which the causal laws did not permit the instantiation of the properties whose instantiation would be involved in its instantiation or in its exercise.

Nothing I have said precludes the possibility of there being worlds in which the causal laws are different from those that prevail in this world. But it seems to follow from my account of property identity that if the laws are different then the properties will have to be different as well. And it does not appear that we have the resources for describing a world in which the properties that can be instantiated differ from what I shall call the 'actual world properties', that is, those that can be instantiated in the actual world. We have just seen that we cannot do this by imagining the conditional powers that exist in this world to be governed by different laws, and so to be differently grouped into properties.

It might seem that we can at least imagine a world in which *some* of the properties that can be instantiated are actual world properties while others are not. But a specification of the causal potentialities of one property will involve mention of other properties, a specification of the causal potentialities of those other properties will involve mention still other properties, and so on. If there could be a world in which some but not all of the actual world properties can be instantiated, this could only be because those properties were causally insulated, as it were, from the rest – that is, were such that their causal potentialities could be fully specified without reference to the rest and vice versa. It seems unlikely that any proper subset of the actual world properties is causally insulated in this way – and any that are insulated from all properties we know about are thereby insulated from our knowledge and our language. But could there be a world in which the properties that can be instantiated include all of the actual world properties plus some others? This would be possible only if the two sets of properties, the actual world properties and the properties that cannot be instantiated in the actual world, were causally insulated from one another. And because of this, it would be impossible for us to say anything about the properties that cannot be instantiated in the actual world; for what we can describe is limited to what can be specified in terms of properties that can be so instantiated. What we could describe of such a world would have to be compatible with the laws that specify the causal potentialities of the actual world properties and, what we have found to be inseparable from these, the laws describing the lawlike connections between the conditional powers that constitute these properties.

Now let us consider the second attempt to reconcile the claim that the identity of a property is determined by its causal potentialities with the apparent conceivability of worlds in which the causal laws that obtain are different from, and incompatible with, those that obtain in the actual world. This involves the proposal that we adopt a 'core cluster theory' in place of the 'total cluster theory', and make the identity of a property depend on a proper subset, rather than on the totality, of the causal potentialities it has in the actual world. Like the first attempted reconciliation, this involves the idea that at least some of the lawlike connections between conditional powers hold only contingently; it is this that is supposed to make it possible for the composition of the total cluster associated with a property to differ from one possible world to another, owing to different conditional powers being causally linked with the conditional powers in the property's essential core cluster. But it would seem that the lawlike connections between those conditional powers included in the essential core cluster will have to hold of logical necessity, i. e., in all possible worlds. For if they held only contingently, then in some possible worlds they would not hold. In such a world, the individual conditional powers which in the actual world constitute the essential core of the property could be instantiated, but the property itself could not be instantiated. Even if these conditional powers could be instantiated together in such a world, their coinstantiation would not count as the instantiation of a property, and so of that property, since the requistite causal unity would be lacking. But I have already argued, in discussing the first attempted reconciliation, that it is not possible that there should be a world in which conditional powers that are instantiated in the actual world can be instantiated while actual world properties cannot be instantiated.

But if, as I have just argued, the lawlike connections between conditional powers within the essential core cluster will have to hold of logical necessity, then we are faced with a problem. Some lawlike connections between conditional powers will hold contingently (according to the core cluster theory), while others will hold as matter of logical necessity. How are we to tell which are which? It does not appear that we can distinguish these lawlike connections epistemologically, i. e., by the way in which they are known. For if, as I am assuming, there are truths that are necessary a posteriori, the fact that a connection is discovered empirically is no guarantee that it does not hold necessarily. Nor can it be said that we identify the necessary connections by the fact that they hold between conditional powers belonging to some property's essential core cluster; for this presupposes that we have some way of identifying essential core clusters, and how are we to do this if we do not already know which connections between conditional powers are necessary and which are contingent?

It might be suggested that what constitutes a set of conditional powers as constituting an essential core cluster is just its being a lawlike truth that whatever has any of its members has all of them, and that it is by discovering such lawlike truths that we identify essential core clusters. Given that the lawlike connections between members of essential core clusters hold of logical necessity, this would amount to the claim that if two conditional powers are so related that the possession of either of them is both causally necessary and causally sufficient for the possession of the other, then the lawlike connection between them holds as a matter of logical necessity, while if the possession of one is causally sufficient but not causally necessary for the possession of the other then the lawlike connection may be contingent. I have no knockdown argument against this view, but it seems to me implausible. If it is possible for it to be a contingent fact that the possession of one conditional power is causally sufficient for the possession of another, then it seems to me that it ought to be possible for it to be a contingent fact that the possession of one conditional power is both causally necessary and causally sufficient for the possession of another; that is, it ought to be possible for it to be contingently true of two conditional powers that the possession of either of them is causally sufficient for the possession of the other. So if we deny that the latter is a possibility, we should also deny that the former is.

It may be suggested that it is our linguistic conventions that make certain causal potentialities essential to a property, and so determine the makeup of a property's essential core cluster. But this cannot be so. It may in some cases belong to the conventionally determined sense of a property word that the property it designates has certain causal potentialities; while I think there is no need for property words to have such Fregean senses, and think that such words often function much as Kripke thinks natural kind terms do, I have no wish to deny that a property word can have a conventionally determined sense. But there is only so much that linguistic conventions can do; and one thing they cannot do is to dictate to reality, creating lawlike connections and *de re* necessities. Having discovered that certain conditional powers necessarily go together, and so are appropriately related for being part of an essential core cluster, we can lay down the convention that a certain word applies, in any possible world, to those and only those things having those conditional powers. But this leaves open the question of how we know that the conditional powers in question are appropriately related – that they must go together in any world in which either can be instantiated. And here appeal to convention cannot help us.

It begins to appear that if we hold that some lawlike connections are contingent, there is no way in which we could discover which of the lawlike connections between conditional powers are logically necessary and which are logically contingent, and so no way in which we could identify the essential core clusters of properties. This means that when we conceive, or seem to be conceiving, of a possible world in which the actual world properties are governed by somewhat different laws, there is no way in which we can discover whether we are conceiving of a genuine possibility. All that any of our empirical investigations can tell us is what lawlike connections obtain in the actual world; and without some way of telling which of these connections are contingent and which necessary, this gives us no information about what can be the case in other possible worlds. This makes all talk about what logically might be and might have been completely idle, except where questions of logical possibility can be settled a priori. If the core cluster theory makes the modal status of causal connections, their being necessary or contingent, epistemologically indeterminate in this way, it does not really save the intuitions which lead us to resist the total cluster theory, according to which all such connections are necessary. Unless we are prepared to abandon altogether the idea that there is a 'fact of the matter' as to whether there are logically possible circumstances in which a given property would make a certain contribution to the causal powers of its subject, I think we must accept the total cluster theory and its initially startling consequence that all of the causal potentialities of a property are essential to it.

X

If, as my theory implies, there are no situations that are logically but not causally possible, how is it that we are apparently able to conceive or imagine such situations? Saul Kripke has suggested one answer to a very similar question.[17] He holds that it is a necessary truth that heat is molecular motion, but recognizes that it seems as if we can imagine heat turning out to be something other than this. According to Kripke, this appearance of conceivability is something to be explained away, and he explains it away by claiming that the seeming conceivability of heat turning out not to be molecular motion consists in the actual conceivability of something else, namely of sensations of a certain sort, those that we in fact get from heat, turning out to be caused by something other than molecular motion. The latter really is conceivable, he holds, and for understandable reasons we mistake its conceivability for the conceivability of something that is in fact not conceivable.

But if conceivability is taken to imply possibility, this account commits one to the possibility that the sensations we get from heat might standardly be caused by something other than molecular motion (and so something other than heat); more than that, it commits one to the possibility that this might be so and that these sensations might be related to other sensations and sense experiences in all the ways they are (or have been to date) in the actual world. And since the property of having such sensations is one that is actualized in this world, this would commit one, in my view, to the claim that it is compatible with the laws of nature that prevail in the actual world that these sensations should be so caused and so related to other experiences. Now this claim may be true – if 'may be' is used epistemically. But it is hard to see how we are entitled to be confident that it is. For might there not be laws, unknown to us, that make it impossible that the standard cause of these sensations should be anything other than it is, given the way they are related to the rest of our experience? If the seeming conceivability of heat turning out to be something other than molecular motion does not prove the actual possibility of this, why should the seeming conceivability of certain sensations being caused by something other than molecular motion prove the actual, and so causal, possibility of that? And if seeming conceivability no more proves possibility in the latter case than in the former, there seems little point in distinguishing between conceivability and seeming conceivability; we may as well allow that it is conceivable (and not just seemingly conceivable) that heat should turn out to be molecular motion, and then acknowledge that conceivability is not conclusive proof of possibility. We could use the term 'conceivable' in such a way that it is conceivable that *P* just in case not-*P* is not provable a priori. Or we could use it in such a way that it is conceivable that *P* just in case it is epistemically possible that it is possible that *P* should be the case – that is, just in case *P*'s being possible is compatible, for all we know, with what we know. These uses of 'conceivable' are not equivalent, but on both of them it is possible to conceive of what is not possible.

XI

Although many of the implications of the account I have advanced are radically at odds with Humean views about causality, it does enable us to salvage one of the central tenets of the Humean view, namely the claim that singular causal statements are 'implicitly general'. As I see it, the generality of causal propositions stems from the generality of properties, that is, from the fact that properties are universals, together with the fact which I began this essay by pointing out, namely that causal relations hold between particular events in virtue of the properties possessed by the constituent objects of those events, and the fact, which I have tried to establish in the paper, that the identity of a property is completely determined by its potential for contributing to the causal powers of the things that have it. If I assert that one event caused another, I imply that the constituent objects of the cause event had properties which always contribute in certain ways to the causal powers of the things that have them, and that the particular episode of causation at hand was an actualization of some of these potentialities. I may of course not know what the relevant properties of the cause event were; and if I do know this, I may know little about their causal potentialities. This is closely related to the now familiar point that in claiming to know the truth of a singular causal statement one is not committed to knowing the laws in virtue of which it holds.[18] Moreover, a singular causal statement does not commit one to the claim that the instantiation of the relevant properties in relevant similar circumstances always produces the effect that it did in the case at hand; for the laws governing these properties may be statistical, the powers to which the properties contribute may, accordingly, be statistical tendencies or propensities, and the causation may be nonnecessitating. Also, the claim that singular causal statements are implicitly general does not, as here interpreted, imply anything about how such statements are known – in particular, it does not imply the Humean view that causal relationships can only be discovered *via* the discovery of regularities or 'constant conjunctions'. But where the present theory differs most radically from theories in the Humean tradition is in what it claims about the modality of the general propositions, the laws, that explain the truth of singular causal propositions; for whereas on the Humean view the truth of these propositions is contingent, on my view it is logically necessary. I thus find myself, in what I once would have regarded as reactionary company, defending the very sort of 'necessary connection' account of causality which Hume is widely applauded for having refuted.[19]

Cornell University

NOTES

[1] See Jaegwon Kim, 'Causation, Nomic Subsumption, and the Concept of Event', *Journal of Philosophy* **70** (1973), 27–36. I should mention that it was reflection on this excellent paper that first led me to the views developed in the present one.

[2] Peter Geach, *God and the Soul*, Routledge and Kegan Paul, London, 1969, p. 71. See also Jaegwon Kim, 'Non-Causal Relations', *Noûs* **8** (1974), 41–52, and 'Events as Property Exemplifications', in M. Brand and D. Walton (eds.), *Action Theory*, Reidel, Dordrecht, 1976, pp. 159–177.

3 I take this example from Kim, 'Causation, Nomic Subsumption, and the Concept of Event'.

4 David Hume, *A Treatise of Human Nature* 1.3.14, ed. by L. A. Selby-Bigge, Oxford University Press, London, 1888, p. 170.

5 "When any objects *resemble* each other, the resemblance will at first strike the eye, or rather the mind, and seldom requires a second examination" (*Treatise* 1.3.1, p. 70).

6 John Locke, *An Essay Concerning Human Understanding*, 2 vols., 2.23.8, ed. by A. C. Fraser, Oxford University Press, London, 1894, Vol. 1, p. 398.

7 *Essay* 2.8.8, Vol. 1, p. 398.

8 What does "in virtue of" mean here? For the moment we can say that a thing has a power in virtue of having certain properties if it is a lawlike truth that whatever has those properties has that power. On the theory I shall be defending it turns out that this is a matter of the possession of the properties entailing the possession of the power (that is, its being true in all possible worlds that whatever has the properties has the power).

9 In speaking of "circumstances" I have in mind the relations of the object to other objects; instead of speaking of "presence in circumstances of a particular sort" I could instead speak of 'possession of particular relational properties'. Being in such and such circumstances is a mere-Cambridge property of an object, not a genuine (intrinsic) property of it.

10 After this was written I found that Peter Achinstein has advanced a causal account of property identity which, despite a different approach, is in some ways similar to the account proposed here. See his 'The Identity of Properties', *American Philosophical Quarterly* **11** (1974), 257–276. There are also similarities, along with important differences, between my views and those presented by D. H. Mellor in 'In Defense of Dispositions', *Philosophical Review* **83** (1974), 157–181, and those presented by R. Harre and E. H. Madden in *Causal Powers: A theory of Natural Necessity*, Blackwell, Oxford, 1975.

11 See. S. F. Barker and P. Achinstein, 'On the New Riddle of Induction', *Philosophical Review* **69** (1960), 512–522. The definition given there is not equivalent to that originally given by Goodman, in *Fact, Fiction and Forecast*, 3rd ed., Bobbs-Merrill, Indianapolis, 1975, p. 74 (originally published by University of London Press, London, 1954), and it is the latter which is employed elsewhere in the present paper.

12 This was called to my attention by Nicholas Sturgeon.

13 See Boyle, 'The Origins and Forms of Qualities', in *The Works of the Honourable Robert Boyle*, 5 vols., A. Millar, London, 1744, Vol. 2, p. 461ff.

14 See Saul Kripke, 'Naming and Necessity', in D. Davidson and G. Harman (eds.), *Semantics of Natural Language*, Reidel, Dordrecht, 1972, pp. 253–355.

15 There is another reason why these conditional powers should not count as constituting a single property: the instantiation of the second of them is causally incompatible with the activation of the first, since something that is such as to be malleable at 100°F cannot be made of steel.

16 For the notion of 'reference fixing', see Kripke, *op. cit.*, pp. 269–275.

17 *Ibid.*, pp. 331–342.

18 See, for example, Donald Davidson, 'Causal Relations', *Journal of Philosophy* **64** (1967), 691–703.

19 Earlier versions of this paper were read at Cornell University, Ithaca, N.Y., in 1975, at the University of Toronto and Temple University, Philadelphia, in 1976, and at the

University of Virginia, Charlottesville, in 1977, and a number of people have given me valuable criticisms of one or more of those versions. Special thanks are due to Richard Boyd, Robert Cummins, Lawrence Davis, Glenn Kessler, Jaegwon Kim, Richard Miller, David Reeve, Michael Slote, Robert Stalnaker, and Peter van Inwagen.

Note Added in Proof. Richard Boyd has offered the following as a counterexample to the account of properties proposed in this paper. Imagine a world in which the basic physical elements include substances *A*, *B*, *C* and *D*. Suppose that *X* is a compound of *A* and *B*, and *Y* is a compound of *C* and *D*. We can suppose that it follows from the laws of nature governing the elements that these two compounds, although composed of different elements, behave exactly alike under all possible circumstances – so that the property of being made of *X* and the property of being made of *Y* share all of their causal potentialities. (This means, among other things, that it follows from the laws that once a portion of *X* or *Y* is formed, it cannot be decomposed into its constituent elements.) It would follow from my account of properties that being made of *X* and being made of *Y* are the same property. And this seems counterintuitive. If, as appears, *X* and *Y* would be different substances, the property of being composed of the one should be different from the property of being composed of the other.

I think that this example does show that my account needs to be revised. I propose the following as a revised account which is still clearly a causal account of properties: for properties *F* and *G* to be identical, it is necessary *both* that *F* and *G* have the same causal potentialities *and* (this is the new requirement) that whatever set of circumstances is sufficient to cause the instantiation of *F* is sufficient to cause the instantiation of *G*, and vice versa. This amounts to saying that properties are individuated by their possible causes as well as by their possible effects. No doubt Boyd's example shows that other things I say in the paper need to be amended.

MYLES BRAND

SIMULTANEOUS CAUSATION

Consider the following claims about causal simultaneity.

(A) All causes are simultaneous with their effects.
(I) Some causes are simultaneous with their effects.
(E) No causes are simultaneous with their effects.

Hume and his followers, including some contemporary philosophers, assert (E). Many philosophers have thought that (I) is the most plausible thesis. In *Action and Purpose*, Richard Taylor argues for (I).[1] In section I, I shall examine these arguments. In Section II, I shall sketch a theory of causal connection that is intended to make (A) plausible. In defending (A), I find myself in agreement with some remarks made by Taylor in some of his earlier writings, and, perhaps, Russell.[2]

If either (I) or (A) is true, then there are important consequences for the analysis of singular causal statements. It is obvious that a true singular causal statement is asymmetrical: if one event causes another, then it is not the case that the latter causes the former. As is well known, Hume attempted to account for this asymmetry by means of temporal succession; this strategy is also followed by some contemporary philosophers, for example, Salmon and Suppes.[3] But the truth of (A) or (I) militates against this approach. Indeed, the truth of (A) or (I) brings to the forefront the problems of accounting for the asymmetry of singular causal statements. In this paper, however, I shall not attempt to provide an account of this asymmetry, nor even survey the most plausible extant accounts.[4] Rather, I shall be concerned with the temporal relationship between cause and effect.

I. TAYLOR'S DEFENSE OF (I)

In *Action and Purpose*, Taylor argues for (I) primarily by arguing against (A) and (E). His argument against (E) consists in presenting some intuitively compelling cases of simultaneous causation. His argument against (A), which is similar to Hume's, is a *reductio*, in that (A) is claimed to be inconsistent with there being causal chains. I shall first discuss the case against (E), and then the *reductio* argument.

Peter van Inwagen (ed.), Time and Cause, 137–153.
Copyright © 1980 *by D. Reidel Publishing Company.*

Consider the following two purported cases of simultaneous causation, the first of which is Taylor's.[5]

I. A locomotive and caboose are attached by a couple. The moving of the locomotive causes the moving of the caboose.

II. Jack and Jill are seated on a seesaw. Jack's going down causes Jill's going up.

These cases rely on two assumptions, which will be accepted without argument. First, they assume that events are the relata of the causal relation. Taylor and others sometimes speak of the relata as conditions or factors; this view can be reconstructed into one saying that the relata are events.[6] Some philosophers have recently urged that causation is a relation between facts or other propositional-like objects.[7] In addition to the persuasive systematic arguments of Davidson against that view,[8] it should be noted that the entire issue of simultaneous causation and temporal succession is senseless if it is not assumed that causation is a relation between events. Events, not abstract entities such as facts, stand in temporal relations. The second assumption is that these events, and the material objects they involve, are the events and material objects of ordinary experience. This assumption precludes taking events to be idealized scientific objects, such as spatiotemporal points. Philosophers of science writing about causation often deny this assumption. But in this paper, I am concerned with causation as an element in our ordinary experience, not in finding a reconstruction of it that would help make sense of some scientific theories. I believe that there are good reasons for taking this approach, for it is through this approach that light can be shed on the philosophical issues involved in action theory, freedom of the will, knowledge, reference, perception, and so on; but I shall not pursue the point here.

Even if these two assumptions are granted, cases I and II are not uncontroversial. It might well be objected that these cases do not occur in the actual world. In case I, the locomotive begins to move slightly before the caboose because the couple stretches, which takes time. Similarly, in case II, Jack starts to go down slightly before Jill begins to go up because the seesaw bends slightly. In both cases, the cause begins slightly before the effect begins. With respect to case I, Taylor replies as follows:

> . . . It is tempting in such an example to suppose that the locomotive must begin moving before the motion of the caboose can begin, but this is not relevant. The effect under consideration is not the caboose's beginning to move but its being in motion.[9]

This reply consists in changing the example. Originally, case I connected the event of the locomotive's movement from start to finish with the event of the caboose's movement from start to finish. But given the elasticity of the couple, the event of the locomotive's movement begins before the event of the caboose's movement. If we divide these events more finely, and consider only the event of the locomotive's moving from the point after the couple has been fully stretched to the point before it begins to compress and, similarly, the event of the caboose's moving from the point after it is fully stretched to the point before it begins to compress, then they appear simultaneous.

Taylor's reply is suggestive of the theory of causal connection I shall develop in the next section. 'The locomotive's moving', 'Jack's going down on the seesaw', and indeed almost all ordinary-language descriptions of causes and effects, name complex events. It is the events contained within these complexes that bear causal relations, not the entire complexes. However, Taylor's reply, as stated, is incomplete and *ad hoc*. For instance, he does not specify how the event of the locomotive's moving while the couple is being stretched fits into the complex situation. In general, he does not present a criterion for dividing these complex events into parts, and does not indicate the relations among these events contained in the complex. Once the underlying theory of causal connection is specified, I contend, it is highly plausible to view all cases of causation as simultaneous.

In his *Monist* article, Taylor says that if an adequate account of case I yields an instance of simultaneous causation, then all cases of causation can be similarly analyzed and seen to be simultaneous.[10] However, in his later writings, especially *Action and Purpose*, he follows Hume in defending the view that only some cases of causation are simultaneous. Taylor argues,

> . . . that *all* causes are contemporaneous with their effects . . . is provably false. For if this were true, then there would be no such thing as a causal chain. Indeed, it would be impossible for any two events whatever to have any causal connection with each other in case there were any lapse of time at all between their occurrences. If some event *A*, for example, causes *B*, which in turn causes *C*, which in turn causes *D*, then in case every cause is simultaneous with its effect, it follows that when *A* occurs, then the others, and indeed every event in the universe that is in any way causally connected with *A*, must occur *at the same time*. This, however, is false. There *are* causal chains, and sometimes temporally separated events are causally related in one way or another.[11]

This argument is reminiscent of a passage in Hume's *Treatise*, which is worth quoting in its entirety not only because its excessiveness is humorous, but also because it suggests a second, different argument that I will discuss later.

> 'Tis an establish'd maxim both in natural and moral philosophy, that an object, which exists for any time in its full perfection without producing another, is not its sole cause;

> but is assisted by some other principle, which pushes it from its state of inactivity, and makes it exert that energy, of which it was secretly possest. Now if any cause may be perfectly co-temporary with its effect, 'tis certain, according to this maxim, that they must all of them be so; since any one of them, which retards its operation for a single moment, exerts not itself at that very individual time, in which it might have operated; and therefore is no proper cause. The consequence of this wou'd be no less than the destruction of that succession of causes, which we observe in the world; and indeed, the utter annihilation of time. For if one cause were co-temporary with its effect, and this with *its* effect, and so on, 'tis plain there wou'd be no such thing as succession, and all objects must be co-existent.[12]

In the next few lines, Hume backs off from this *reductio* argument, and begs the reader to continue even if the argument is not satisfactory. Hume did well not to press the argument, since it is unsound. Let us look carefully at Taylor's version of it, which can be reformulated as follows.

(1) All causes are simultaneous with their effects (assumption).
(2) If all causes are simultaneous with their effects, then it is impossible that there are two causally connected events that are not simultaneous.
(3) If it is impossible that there are two causally connected events that are not simultaneous, then there are no causally connected but temporally separated events.
(4) There are causal chains only if there are causally connected but temporally separated events.
(5) There are causal chains.

Therefore, from (4) and (5)

(6) There are causally connected but temporally separated events.

Therefore, from (1), (2), (3), and (6)

(7) It is not the case that all causes are simultaneous with their effects.

There is slippage between 'causes' and 'causally connected' and there is the interjection of impossibility in premises (2) and (3). But these defects are harmless: interpret 'causal connection' to be restricted to cause and effect; and drop the reference to what is impossible, replacing throughout 'it is not the case that' for 'it is impossible that'. Premise (4) is not explicitly stated by Taylor, but it appears assumed, and in any case, is required for the validity of the argument. (Similarly, in Hume's version of the argument, it is assumed that there is temporal succession only if causally connected events are not co-temporal.) This premise, however, is false; and hence, the *reductio* of (1) fails.

The existence of causal chains does not require that there is a temporal lag between the beginning of the cause and the beginning of the effect. Some causal chains consist in sequences of events, each event in a sequence temporally related to others in that sequence, and initial and terminal members of each sequence causally related and simultaneous with other terminal and initial members of other sequences. Diagrams might help here. Premise (4) is true if Figure 1 obtains, where $e, f, \ldots$, are events, and where causal connections are depicted by horizontal contact.

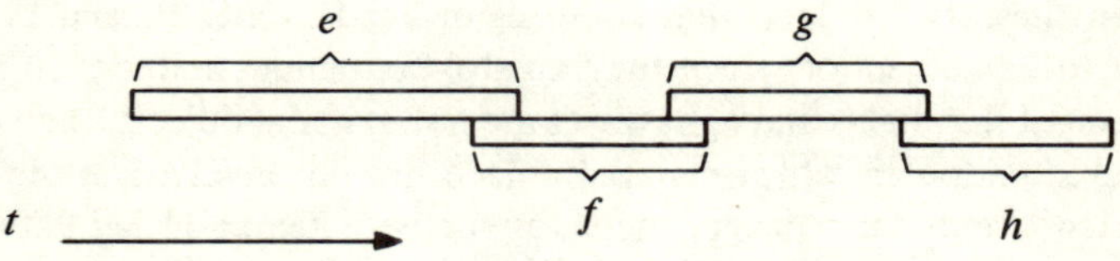

Fig. 1.

But (4) is false if Figure 2 pictures what takes place at least sometimes.

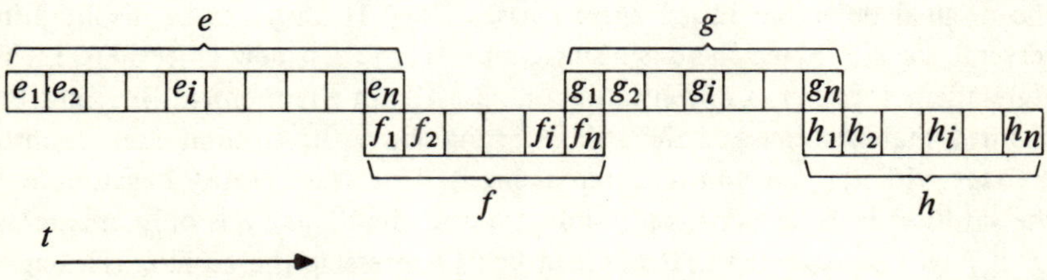

Fig. 2.

In Figure 2, $e, f, \ldots$, are taken to consist of sequences of events, $e_1, \ldots, e_i, \ldots, e_n$; $f_1, \ldots, f_i, \ldots, f_n$; and so on, respectively; the events in these sequences are not causally related. The terminal event of one sequence causes the initial event of another, and these two are simultaneous. Figure 2 depicts causally connected simultaneous events occurring in causal chains. In the next section I shall argue for the plausibility of viewing many causal situations in the way illustrated by Figure 2.

II. ALL CAUSES ARE SIMULTANEOUS WITH THEIR EFFECTS

Let me multiply purported cases of simultaneous causation, the first of which is based on an example of Kant's.

III The lead ball's going down causes the depressing of the cushion.
IV Jack's going down on the seesaw causes the seesaw's going down.
V The increasing of the gravitational field casuses the bending of the light beam.

It was objected to cases I and II that there is a temporal lag between movement of one material object – the locomotive or Jack – and the movement of another material object – the caboose or Jill, since there is a connection between these two objects that stretches or bends. Cases III and IV are not subject to this objection, since they involve contiguous material objects.[13] If it is granted that fields and light waves are not material objects, then case V is a causal situation in which material objects are not involved in the causally connected events; and hence, there cannot be a temporal lag between the movements of the material objects.[14] In reconstructing case I into one in which the couple between the locomotive and caboose is fully stretched, Taylor essentially changed to a situation in which the material objects are contiguous. A similar change in case II can be made, in which the movements are considered after the seesaw reaches maximal curvature.

Although cases III – V, and cases I and II reconstructed, are not subject to the original objection raised against cases I and II, they can be disputed in serveral ways.[15] Since there is a single rebuttal to these new objections, I will state them together. Concentrating on case III for the moment, it might be claimed that the cause is the ball's starting above the cushion, coming into contact with the cushion and depressing it. This event clearly began before the cushion became depressed, and thus the simultaneity is only apparent. Or it might be objected that the lead ball's depressing the cushion is a single event involving two material objects. There is some other event, not named in the report of the case, that brought the ball in contact with the cushion, and it is that event that caused the lead ball's depressing the cushion. Both these objections are based on demarcating events differently from the way they are demarcated on a first reading of the cases. Other objections can be generated by demarcating the events in still other ways.

As I indicated above, I shall assume that events are the relata of the causal relation. But something must now be said about the nature of events. I have argued in several other papers that events are particulars, in that they are spatiotemporally locatable and not literally repeatable.[16] Events are similar to material objects, the essential difference being that two events can occupy a single spatiotemporal region, whereas two material objects cannot. For example, the events of Esther Williams swimming the English Channel and Esther Williams catching a cold can be spatiotemporally coincidental, as can the events of its raining in Chicago at 6.00 P.M. yesterday and its growing dark in Chicago at 6.00 P. M. yesterday.

The nature of a kind of object is specified by means of identity conditions. Letting $e, f, \ldots,$ range over events; $o_1, o_2, \ldots,$ over material objects; s over spatiotemporal regions, and letting R be the relation '... occurs within — ', the identity conditions for material objects and events are the following (assuming closure):

(I1) $o_1 = o_2$ iff $(s)\,(o_1Rs \equiv o_2Rs)$;

(I2) $e = f$ iff $\Box\,(s)\,(eRs \equiv fRs)$.

That is, o_1 and o_2 name a single material object just in case they name something occupying the same spatiotemporal regions; and e and f name a single event just in case they name something necessarily occupying the same spatiotemporal regions.

It is important to observe that (I1) and (I2) are schemas. They do not indicate which descriptions or names of events and material objects can be substituted into (I1) and (I2). Restricting attention to (I2), let us follow Kim and take 'x's F-ing in spatiotemporal region s' to be the canonical form for describing (monadic) events where x usually but not always names a material object.[17] In order to apply the identity conditions (I2) to some actual cases, it must be modified. For example, consider these two descriptions of a single event: 'Walter Mondale's giving a speech in Boise now' and 'the Vice-President of the United States giving a speech in Boise now'. Since 'the Vice-President of the United States' names someone other than Mondale in some possible worlds, to use the common informal semantical idiom, the right side of (I2) is false and the left-hand side is true. My solution to this and related problems of application is to freeze the reference to the actual world to the *contained* names and descriptions in the complex event descriptions e and f. For this purpose, we can use Kaplan's *Dthat* operator, which makes these names and descriptions into rigid designators naming in every possible world in which they exist the objects named in this world. Letting e^+ be the result of *Dthat*-ing the contained names and descriptions in e, we obtain the following version of (I2):

(I2$^+$) $e = f$ iff $\Box\,(s)\,(e^+Rs \equiv f^+Rs)$.

This version is adequate to the above case: 'the Vice-President of the United States^{+}' names Mondale in every possible world (in which he exists), and hence, 'Walter Mondale's giving a speech in Boise (now)' and 'the Vice-President of the United States giving a speech in Boise (now)$^{+}$' name a single event. It is important to note that ($I2^{+}$) is not able to determine whether two descriptions name a single event when the descriptions are not in canonical form. For instance, ($I2^{+}$) does not tell us whether 'Hurricane Alice' and 'my least favorite event' name one or two events. But this is as it should be, since these descriptions do not provide sufficient information for this determinations; background information about my least favorite event is required. In order to simplify the discussion about simultaneous causation, I shall use descriptions of events that can be accommodated by (I2). The main point, to repeat, is that identity conditions for events are in terms of necessary spatiotemporal coincidence.

Returning to the first objection against case III, it was claimed that the cause and effect are not simultaneous because the cause consisted in the lead ball's starting above the cushion, coming into contact with it, depressing it, and coming to rest, whereas the effect coincided temporally only with the ball's depressing of the cushion until it came to rest. The lead ball's starting above the cushion, its coming into contact with it, its moving downward on the cushion and its coming to rest are all different events. For example, it is logically possible for the lead ball to occupy the spatiotemporal regions of starting above the cushion and its falling without its also occupying the spatiotemporal regions of its depressing the cushion or of its coming to rest on the cushion. The cause here is the lead ball's coming into contact with the cushion and the effect is the cushion's beginning to be depressed. The lead ball's coming into contact with the cushion is the last part of a sequence of events beginning with the ball's starting to fall, and the cushion's beginning to be depressed is the first part of a sequence of events ending with the cushion's being fully depressed. It is the last event of the former sequence and the first event of the latter one that are causally related and these events are simultaneous. Similarly, (I2) yields an answer to the second objection, that there was a single event of the ball's moving down on the cushion which was caused by a prior event. The ball's moving down and the cushion's becoming depressed are distinct, since the spatiotemporal regions occupied by each are distinct. Other objections that are based on dividing events in case III can be dealt with in a similar manner.

Let us return to cases I and II. The original objection to these cases is that the seesaw bends slightly, and hence Jack moves before Jill starts to move, and that the couple stretches slightly, and hence the locomotive moves before the caboose starts to move. Considering case II, Jill's going up is not a single event, but rather a sequence of events, the first of which is Jill's beginning to go up. Similarly, the seesaw's bending is a sequence of events, the last of which is its reaching maximal curvature and the first of which is its beginning to curve. Likewise, Jack's going down is a sequence of events, the first of which is his beginning to depress the seesaw. Case II initially appeared to be an instance of two causally related simultaneous events; but counting events on the basis of (I2) shows that there are three sequences of events, the terminal event of the first sequence causing the initial event of the second sequence and the terminal event of the second sequence causing the initial event of the third sequence. There are, in short, two causal connections in case II, and each is an instance of simultaneous causation. Case I is to be analyzed in an analogous way.

It is not clear how many spatiotemporal events there are in the sequence of, say, Jill's going up. However, for our purposes, it is not necessary to provide a criterion for dividing events into spatiotemporal parts. All that needs to be said is that events are happenings in the world, and, hence, cannot be idealized objects, such as spatiotemporal points. Events, it is important to note, may be divided into parts other than spatiotemporally. Suppose, for example, I vote on a particular occasion by flipping the lever. Here my flipping the lever is part of the entire action, but not merely a spatiotemporal part. Rather, my flipping the lever conventionally generates (to use Goldman's terminology) my casting the vote. It is an important and central issue, both for event theory and action theory, how events are to be divided into parts and how the parts are related to wholes.[18] But in whatever way this complex issue is resolved, the following must be true: events such as Jill's going up on the seesaw are divisible into spatiotemporal parts that are themselves events.

At this point, I should like to consider another kind of case:

VI From 5.00 to 5.30, there is a pork chop on a plate on the dinning-room table. The table's remaining the way it is causes the plate's remaining stationary, and the plate's remaining stationary causes the pork chop's remaining stationary.

This is a case of equilibrium, in which the material objects involved in the events are caused to remain unchanged during a time interval. Equilibrium cases, like the others considered, consist of sequences of events. The table's remaining stationary during the interval 5.00 to 5.30, for instance, is a sequence of spatiotemporal events (though it is nuclear how many events there are in that sequence). Each event in this sequence causes an event in the sequence of the plate's remaining stationary, and each event in the sequence of the pork chop's remaining stationary. This situation is illustrated in Figure 3, where, again, horizontal contact represents causality:

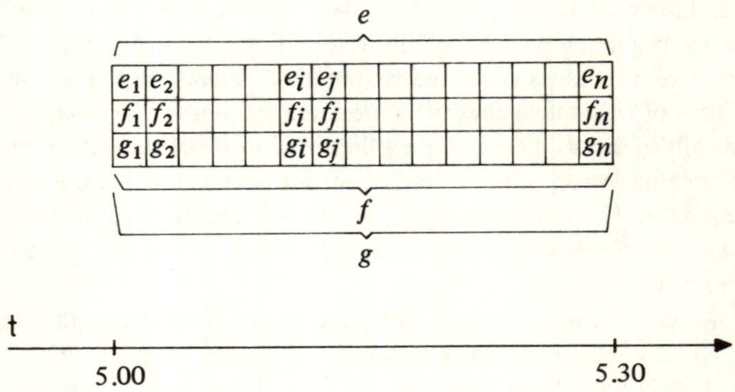

Fig. 3.

If the floor's keeping the table stationary were taken as part of this example, then another sequence, say *h*, consisting of $h_1, h_2, \ldots, h_n$, would be depicted directly below *g*.

Case VI is a situation involving *n* causal chains, each consisting of three links. Thus not all causal chains contain temporally sequential events; some causal chains contain simultaneous events. Notice that the equilibrium cases also show that (4) of Taylor's *reductio* argument is false. These causal chains, like those containing temporally sequential events, do not consist of temporally separated, causally connected events.

It is convenient to introduce the following terminology. Let a *causal interaction* be a situation in which one event causes another; and let a *causal process* be a sequence of events such that the initial and terminal events in the sequence, and only these events, are involved in causal interactions.[19] Event sequences can be temporally extended or simultaneous; and hence, causal processes can be temporally extended or simultaneous. A *causal chain* consists of at least two causal processes such that the terminal event of each process causally interacts with the initial event of another process. Using this terminology, case II, for instance, involves a causal chain (as illustrated by Figure 2) consisting of four temporally extended causal processes. Case VI, the equilibrium example, involves as many causal chains as there are spatiotemporal eventual parts in each sequence. The general thesis being defended can be restated by saying that all causal interactions are simultaneous and that temporal succession is possible, since some causal chains contain temporally sequential events.

It might now be objected that the problem of simultaneous causation has only been pushed back a step. What reason is there to maintain that causal interactions are cotemporal? In the nonequilibrium example, case II, perhaps the seesaw reaches maximal curvature, *and then* Jill begins to rise.

Let *a* be the seesaw's reaching maximal curvature and *b* be Jill's beginning to rise. I have urged that Figure 4 illustrates the actual situation; the objection suggests that Figure 5 or Figure 6 give the correct picture.

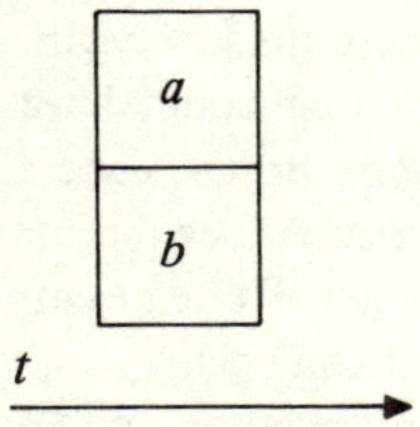

Fig. 4.

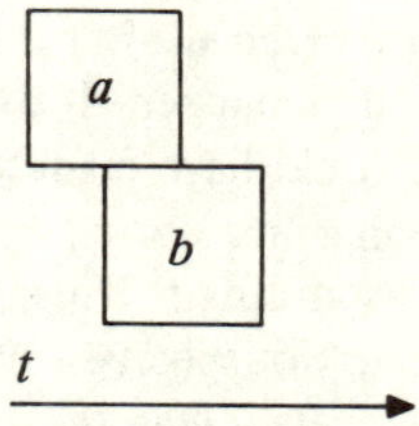

Fig. 5

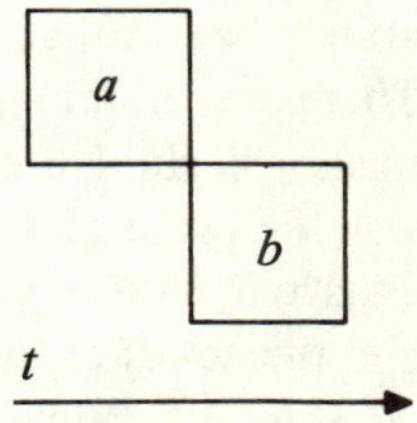

Fig. 6.

The situations depicted by Figure 5 and Figure 6, however, are precluded by means of a principle cited by Hume in his regress argument. Although that argument is unsound, since simultaneous causation is consistent with temporal succession, the maxim used in the argument is true. Recall that Hume says " 'Tis an establish'd maxim both in natural and moral philosophy, that an object, which exists for any time in its full perfection without producing another, is not its sole cause". A natural interpretation is

> *Hume's maxim*. For any events *e* and *f* and time interval *t*, if *e* occurs during *t* and *e* does not change during *t* and *e* is the cause of *f*, then *f* occurs during *t*.

I have rendered "an object . . . exists for [a] . . . time in its full perfection" as 'event *e* does not change during *t*'.

Events can change genuinely or nongenuinely. The latter, so-called Cambridge changes, consist in an event's gaining or losing a relational property.[20] For example, the event of Socrates drinking hemlock might gain the property of being admired by me. But that property seems not to affect the event itself; it seems not to concern the internal structure, as it were, of the event of Socrates drinking hemlock. Some events genuinely change. They acquire or lose a nonrelational property during the time that they occur but remain the same event. For example, a thunderstorm might dissipate, a World Series might become lopsided, a battle might suddenly be lost, or a marriage might be saved. Nothing, in principle, seems to preclude any event from changing relationally; but only some events can genuinely change. Events that have other events as parts might change genuinely, while events not having eventual parts appear not to be subject to genuine change.

Hume's maxim says that if an event *e* occurs during *t* and does not *genuinely* change during *t*, then *f* also occurs during *t* if *e* is the cause of *f*. Suppose that, contrary to the maxim, *e* occurred during *t, e* remained the same during *t, e* was the cause of *f*, but *f* began after *e* began. However, if *f* began after *e* began, then something happened after *e* began to have made *f* begin. There was nothing, however, that happened after *e* began that could have made *f* begin. No other event could have made *f* begin after *e* began, since *e* was the cause of *f*; and no change in *e* during *t* could have made *f* begin after *e* started, since *e* did not change during *t*. Thus, *f* did not begin after *e* began, and the denial of the maxim does not specify a situation that can occur.

Thus, by Hume's maxim, events *a* (the seesaw's reaching maximal curvature) and *b* (Jill's beginning to rise) are simultaneous; for *a* is the cause of *b* and *a* is an unchanging event. If it is thought that, say, *a* changes during the time interval in which it occurs, then there would be a spatiotemporal region in which *a* occurs before the change and a spatiotemporal region in which *a* occurs after the change. By (I2), *a* could then be divided into two events, a_1 and a_2. Provided that a_2 is not further divisible, it is now a_2, whatever exactly it is, that causes *b*. Similarly, if *b* changed, it would be that initial eventual part of *b* that does not change which causally interacts with a_2.

The general line of argument can be summarized as follows. Identity conditions for events (I2) provide a way to restrict the initial and terminal events in a sequence to those that do not genuinely change. Since the initial and terminal events in sequences causally interact, it follows by Hume's maxim that they are simultaneous.

Let us look somewhat more carefully at the interrelations among events in a temporally extended causal process. An example of this type of causal process is Jill's going up, which consists of Jill's beginning to rise, Jill's rising part way, Jill's rising another part of the way, and so on, including Jill's coming to the top. The events in this sequence are not causally connected to each other. It might be helpful to consider another example to see this point. Suppose that a lead ball is dropped from a certain height and falls freely until it hits the ground. The ball's falling is a causal process, the first member of which is its beginning to fall and the last member of which is its coming into contact with the ground. The falling of the ball for, say, the *i*th segment does not cause it to fall for the *j*th segment. Hume's maxim yields the result that these two events are not causally related. Since its falling during the *j*th segment is later than its falling during the *i*th segment, and (agreeing for simplicity) the event of its falling during the *i*th segment does not change during the interval in which it occurs, it follows that its falling for the *i*th segment does not cause its falling for the *j*th segment.

Events in this kind of causal process are temporally odered, but the relationship between them is stronger than that. The events of my falling asleep at midnight and its raining in Newark slightly later than midnight are also temporally ordered. But the former case supports a corresponding counterfactual, whereas the latter does not. That is,

(i) if the ball had not fallen the *i*th segment, it would not have fallen the *j*th segment

is true, but

(ii) if I had not fallen asleep, it would not have rained in Newark

is false.

It might be thought that since (i) is true, the connection between the ball's falling the *i*th segment and its falling the *j*th segment is really causal. But that line of argument is unsound. Counterfactuals are supported by statements indicating various kinds of noncausal connections between events. Consider, for example, Thomas Reid's classic case of day following night. The counterfactual

(iii) if the coming of this night had not occurred, then the coming of this day would not have occurred

is true, where the coming of this day and the coming of this night are related by correlation, not causation. Similarly, consider the true counterfactual

(iv) if Jones had not raised his arm in order to attract the chairman's attention, he would not have raised his arm,

where the events of Jones raising his arm in order to attract the chairman's attention and his raising his arm are connected in some complicated part-whole or conventional way.

The relationship between the ball's falling the *i*th segment and its falling the *j*th segment is stronger than mere temporal succession, but weaker than causality. The relationship appears to depend on at least two factors. First, the events in the sequence involve the same material objects. At least, I cannot think of an uncontroversial example in which the events are temporally but not causally ordered, involve different material objects and the corresponding counterfactual is true. Second, there is at least one background or standing condition, beginning before the sequence began and continuing after it ends. In the case of the freely falling ball, the standing condition is the force of gravity. An understanding of standing conditions is needed for a full understanding of causality; unfortunately, little helpful work has been done on this notion. It is not clear, for example, what is the ontological status of standing conditions. There is something odd in taking them to be events, though, on final analysis, that might be acceptable. Standing conditions, I take it, are causally necessary but not sufficient for the occurrence of an event. The force of gravity, then, is necessary but not sufficient for the ball's falling the *j*th segment. Note it follows that the force of gravity does not cause the ball's falling the *j*th segment, since the cause, whatever else it might be, is sufficient for the effect. The relationship among members in a temporally extended sequence of events can be thought of as a form of physical necessity, though with a warning not to confuse it with causal necessity. Instances of this form of physical necessity exemplify laws of succession. An interesting exercise would be to give a possible-world or other semantics for statements of physical necessity and compare and contrast that semantics with one for causal necessity.

Our ordinary way of talking about causality does not correspond with the analysis of causal situations given here. 'Jack's going down on the seesaw causes Jill's going up' suggests that there are two causally related events, though, as we have seen, upon analysis, this case involves many events and three causal processes. Usually, in our ordinary talk about causality, we rarely name the proximate cause. We might say that Harry's drinking that last martini caused his automobile accident, though Harry finished drinking the martini some time before he entered the car. Our ordinary talk provides summaries of causal situations, often by picking out remote causal processes that are appropriate to the context. It would be tedious and pedantic, at best, or misleading, at worst, to describe the details of the situation in an ordinary setting. I am certainly not suggesting that we abandon our usual way of talking about causality; indeed, for purposes of clarity, I myself have used this way of talking a number of times. Rather, we can continue to talk in the ordinary way provided that we realize that an analysis of these situations yields the result that there are causal processes, the initial and terminal events of each being involved in simultaneous causal interactions. Berkeley's advice on another matter is applicable here: we can speak with the vulgar, though we should think with the learned.

Let me mention one final issue, namely, universal causal determinism. This view is usually formulated as follows:

(D) All events are caused.

I want to observe that if the foregoing analysis of causal connection is correct, there seems to be an easy disproof of (D). An event within a temporally extended causal process that is neither a terminal nor initial event is not caused. However, this result occurs because (D) is formulated using the ordinary way of talking about events and causes. In order to see whether the above account of causal connection really shows that causal determinism is false, the view must be formulated in the terminology introduced. The most natural rendering seems to be

(D′) Every event is part of some causal process.

Recall that the initial and terminal events in a causal process causally interact with events from other causal processes. Hence, (D′) says that every event is involved in a causal interaction or is part of a sequence that begins and ends in a causal interaction. Events that are neither initial nor terminal segments of a causal process nevertheless occur lawfully; for they fall under the laws of succession. Basically, (D′) says that, although not every event is caused, every event is determined. I shall not rehearse here the arguments for or against universal causal determinism; they can all be reformulated, without loss of import, as defending or attacking (D′).[21]

University of Illinois at Chicago Circle

NOTES

[1] Richard Taylor, *Action and Purpose*, Prentice-Hall, Englewood Cliffs, N.J., 1966, pp. 9–39, esp. pp. 35–39.

[2] Richard Taylor, 'Causation', *Monist* 47 (1963), 287–313, esp. pp. 303–313, reprinted in Myles Brand (ed.), *The Nature of Causation*, University of Illinois Press, Urbana, 1976, pp. 277–305; and Bertrand Russell, 'On the Notion of Cause', in *Mysticism and Logic*, Doubleday, Garden City, N.Y., 1917.

[3] Wesley Salmon, 'An "At-At" Theory of Causal Influence', *Philosophy of Science* **44**

(1977), 215–224; Patrick Suppes, *A Probabilistic Theory of Causality*, North-Holland Publishing, Amsterdam, 1970.

4 For reviews of historical and contemporary attempts to account for causal asymmetry, see Myles Brand, 'Causality', in P. Asquith and H. E. Kyburg, Jr. (eds.), *Current Research in Philosophy of Science*, Philosophy of Science Association, East Lansing, Mich., 1979; and Richard Taylor, 'Causation' in Paul Edwards (ed.), *The Encyclopedia of Philosophy*, 8 vols., Macmillan and Free Press, New York, 1967, Vol. 2, pp. 56–66.

5 *Action and Purpose*, pp. 35–36.

6 See, for example, Taylor's discussions in *Monist*, and *Action and Purpose*; and J. L. Mackie, 'Causes and Conditions', *American Philosophical Quarterly* **2** (1965), 245–264, reprinted, among other places, in Brand (ed.), *op. cit.*, pp. 307–344. *Cf.* Jaegwon Kim, 'Causes and Events: Mackie on Causation', *Journal of Philosophy* **68** (1971), 426–441.

7 For a survey of these views, see Mackie, *The Cement of the Universe*, Oxford University Press, London, 1974, Chap. 10.

8 Donald Davidson, 'Causal Relations', *Journal of Philosophy* **64** (1967), 691–703, reprinted, among other places, in Brand (ed.), *op. cit.*, pp. 354–367; and 'The Individuation of Events', in N. Rescher *et al.* (eds.), *Essay in Honor of Carl G. Hempel*, Reidel, Dordrecht, 1970, pp. 216–234.

9 *Encyclopedia of Philosophy*, p. 65; Taylor makes the same point in *Monist*, p. 309 and *Action and Purpose*, pp. 35–36.

10 *Monist*, p. 311.

11 *Action and Purpose*, p. 38.

12 *A Treatise of Human Nature* 1.3.2, ed. by L. A. Selby-Bigge, Oxford University Press, London, 1888, p. 76. For an interesting interpretation of this argument, see Stanley Munsat, 'Hume's Argument That Causes Must Precede Their Effects', *Philosophical Studies* **22** (1971), 24–26.

13 Taylor uses a similar example. Suppose that a person is holding a pencil, then the movement of his hand causes the movement of the pencil. See: *Action and Purpose*, p. 36; *Encyclopedia of Philosophy*, p. 65; *Monist*, pp. 309–310.

14 It is difficult to see how, for instance, an increasing gravitational field can be construed as involving material objects. The cause of such an event might well involve material objects, but the event of the field's increasing in strength appears devoid of material objects. If events of this sort do not involve material objects, then the view that all events involve material objects, held by Kim and others, is mistaken – *cf.* Jaegwon Kim, 'Events as Property Exemplifications', in M. Brand and D. Walton (eds.), *Action Theory*, Reidel, Dordrecht, 1976, pp. 159–177.

15 *Cf.* Alexander Rosenberg, 'Propter Hoc, Ergo post Hoc', *American Philosophical Quarterly* **12** (1975), 245–254, esp. pp. 252–253.

16 Brand, 'Particulars, Events and Actions', in Brand and Walton, *op. cit.*, pp. 135–156; and 'Identity Conditions for Events', *American Philosophical Quarterly* **14** (1977), 329–337.

17 Note that describing events using this canonical form does not commit us to Kim's view of the internal structure of events. (Actually, Kim's canonical form is not exactly the one used here; he includes a temporal specification, but not a spatial one.)

18 See Alvin Goldman, *A Theory of Human Action*, Prentice-Hall, Englewood Cliffs, N.J., 1970. Judith Jarvis Thomson's *Acts and Other Events*, Cornell University Press, Ithaca, N.Y., 1977, contains an extended, original and provocative discussion on parts of events.

[19] Salmon, *op. cit*., follows Hans Reichenbach in *The Direction of Time*, ed. by Maria Reichenbach, University of California Press, Berkeley, 1956, esp. Sections 22–23, in using the terms 'causal interaction' and 'causal process'. There is considerable difference between my use of the terms and Salmon's and Reichenbach's, though there are also some underlying affinities. Some of the differences stem from different notions of events.

[20] For a discussion of Cambridge changes, see Kim, 'Noncausal Connection', *Noûs* 8 (1974), 41–52.

[21] I have found helpful the comments of Ralf Meerbote, Marshall Swain, Paul Teller and Irving Thalberg on an early draft. I have also profited by discussing these issues with a number of my colleagues, especially Daniel Berger, and with members of the audiences at the University of Cincinnati, Ohio, and the Somona State Action Theory Conference.

MARSHALL SWAIN

CAUSATION AND DISTINCT EVENTS

If a specific event *c* can properly be called a cause of another specific event *e*, then *c* and *e* must be *distinct* events. This general principle seems agreed upon by virtually everyone who works on causation. Moreover, it seems generally agreed, or at least assumed, that 'distinct' is synonymous with 'not identical'; indeed, the above principle is often put by saying that no event can be a cause of *itself*. Recent work[1] has suggested, however, that there are various pairs of events (such as 'Cambridge events') which are distinct in the sense of not being identical, but which are such that they cannot be causally related. Moreover, these pairs of events appear to be counterexamples to a number of the main approaches to defining causation. In this paper I want to suggest that all of these problematic cases can be dealt with in an intuitively straight-forward and appealing way. The principle that only distinct events can be causally related needs to be understood in a much broader way than the way in which it is usually understood. Specifically, I shall argue that this principle, to be fully adequate, requires a notion of distinctness of events which is broader than the notion of nonidentity. Once this broad notion of distinctness has been outlined, it will be seen that the troublesome cases are excluded by the principle in question.

I am concerned, in particular, to show how the *counterfactual* analysis of causation can be formulated to avoid these troublesome cases.[2] I believe, however, that the broad notion of distinctness of events that I shall suggest could be adopted by anyone wishing to defend some other approach to causation.

I. THE COUNTERFACTUAL ANALYSIS OF CAUSATION

In this section I shall introduce the basic elements of the counterfactual approach to the analysis of causation. As I shall present it, the approach is concerned to provide an analysis of the locution '*c* is a cause of *e*', where '*c*' and '*e*' are names of specific events that occurred, and where the kind of causation involved is what we usually call efficient causation. There are many other causal locutions to be analyzed. For example, there is the notion of agent causation expressed in locutions such as '*S* was a cause of *e*', where '*S*'

Peter van Inwagen (ed.), Time and Cause, 155–169.
Copyright © 1980 *by D. Reidel Publishing Company.*

is the name of some agent and 'e' is some specific event. There are also various causal expressions in which some event or agent is asserted to be unique in a specified causal role; for example, the locution 'c is *the* cause of e', or the locution 'c was the principal cause of e'. In addition, there are various locutions that express causal generalizations, such as 'Continual stress causes heart attacks'. Even though the counterfactual analysis that I shall be working with in this paper does not address these locutions, I believe that the locution 'c is a cause of e' is central to the analysis of each of them.

In presenting the elements of the analysis I shall use counterfactual expressions. I shall also assume that some version of the 'possible worlds approach' to the theory of counterfactuals is correct, although I do not assume that the success or failure of the counterfactual analysis of causation ultimately hangs on the success or failure of that approach to counterfactuals. The possible worlds approach to counterfactuals has considerable intuitive appeal, and I also find it to have heuristic value when talking about causation.

For the purposes of this paper, I shall take the three-place relation expressed by

(1) World w_1 is closer to world w than is world w_2

as unanalyzed, or 'primitive'. This notion is at the heart of the possible worlds approach to counterfactuals, and ultimately one must come to grips with the problem of the criteria in accordance with which judgments about closeness are to be rendered. According to David Lewis, for example, a world w_1 is closer to a world w than is the world w_2 if w_1 resembles w in respect of *overall comparative similarity* more than w_2 resembles w.[3] But this particular suggestion faces difficult problems, and viable competitors are available.[4] Even with (1) taken as unanalyzed, however, we can state truth conditions for counterfactual expressions which are intuitively natural and appealing. Following Lewis, let us say, where A and C are any two propositions, and where A-worlds and C-worlds are worlds in which A and C hold, respectively:

(D1) The proposition that if A were true, then C would be true is true at a world w iff either
  (1) there are no possible A-worlds, or
  (2) some A-world where C holds is closer to w than is any A-world where C does not hold.

(D1) is formulated in such a way that the world w, as well as the A-worlds and C-worlds considered, need not be identical with our actual world. From this point on, unless otherwise specified, I shall assume that the counterfactuals used *are* being evaluated at the actual world.

A key notion for our purposes is the notion of *counterfactual dependence* between two propositions. A proposition C depends counterfactually on a proposition A if the question whether C is true or not depends upon whether A is true or not. More formally:

(D2) C depends counterfactually on A iff:
(1) if A were true, then C would be true; and
(2) if A were not true, then C would not be true.

Now I shall assume (what is in fact a consequence of Lewis's formal constraints on closeness of worlds) that if A and C are both false, then the counterfactual (2) in (D2) is automatically true; in that case, the counterfactual dependence of C on A reduces to the truth of counterfactual (1) of (D2). Similarly, I shall assume, if A and C are both true, then the counterfactual (1) of (D2) is automatically true; in that case, the counterfactual dependence of C on A reduces to the truth of counterfactual (2).

Counterfactual dependence is a relation that holds, as defined in (D2), between propositions. We can also introduce a relation of counterfactual dependence between *events*. The event e is counterfactually dependent upon the event c if the proposition that e occurs is counterfactually dependent upon the proposition that c occurs. If c and e are *occurrent* events, the proposition that e occurs and the proposition that c occurs will be true. Hence, we can use (D2), and the considerations of the preceding paragraph, to get the following:

(D3) Where c and e are occurrent events, e depends counterfactually on c iff: if c had not occurred, then e would not have occurred.

It is tempting to think that if an event e is counterfactually dependent upon an event c, then e is *causally* dependent upon c. While this is often the case, there are exceptions that must be accounted for; indeed, some of these exceptions lead to the problems that are our primary concern in this paper. The most obvious problem is that there is nothing in (D3) that tells us that c and e cannot be the *same* event. Moreover, it seems quite trivially clear that if c and e are identical, e depends counterfactually on c. So, if we were to take counterfactual dependence to be sufficient for causal dependence, we would get the unhappy result that every occurrent event is dependent causally upon itself. As we noted at the beginning of this paper, there is a requirement generally agreed upon that if one event is to be a cause of another, then the two events must be distinct. The same requirement holds for causal dependence. We can, therefore, transform (D3) into a definition of *causal* dependence by adding a requirement to the effect that the events *c* and *e* are *distinct* events. But then we are faced with the question, when are two events distinct? If our only motivation is to avoid the obvious problem considered above, then the answer seems easy enough:

(D4) Two events, *c* and *e*, are distinct iff *c* and *e* are not identical events.

If (D4) is to be of any service, however, we should provide some criterion for the identity of events. This is, of course, one of the knottiest of problems. There are two main approaches to be considered. One of these results in what Alvin Goldman calls the "fine-grained" view of events,[5] and we may take the other to be the 'coarse-grained' view. Perhaps an example will help to make the difference clear. Consider the following two events,[6] which happen to be actions of a person *S*:

(a1) *S*' s turning on the light;
(a2) *S*'s alerting the prowler.

According to the 'fine-grained' view, these events are not identical, and according to the 'coarse-grained' view, they are identical. There are, in my opinion, powerful and conclusive arguments against the 'coarse-grained' view. Many of these arguments have been presented with admirable clarity by Goldman. I shall not repeat these arguments here, but will assume that some version of a 'fine-grained' view is to be preferred.[7] This means, among other things, that our criterion for event identity should yield the result that a1 and a2, and similar pairs, are nonidentical events.

What shall our criterion for event identity be? Many suggestions have been made. Some of them, while most interesting, would clearly not serve my purposes in this paper. For example, one might provide a criterion for event identity in terms of cause and effect: two events are identical, we might say, just in case they have precisely the same causes and effects. It is not clear, however, that this approach will yield a 'fine-grained' view. Even more importantly, if I were to adopt such a criterion in the present context, then I would be using the notion of causation in analyzing a key factor in my analysis of causation, and thus the analysis would be circular. From among the many possibilities for a criterion of event identity, I shall adopt the following:[8]

(D5) Two events, *c* and *e*, are identical iff, necessarily, *c* occurs if and only if *e* occurs.

The kind of necessity intended in (D5) is logical necessity. If we adopt a possible worlds semantics for this modal notion, then our criterion says that two events are identical if and only if they occur in all the same possible worlds. Two events will thus be nonidentical just in case there is at least one possible world in which one of them occurs and the other does not. We can see that this criterion yields the nonidentity of a1 and a2, for there surely are possible worlds in which *S* turns on the light without alerting the prowler, and there are possible worlds in which *S* alerts the prowler without turning on the light.[9] In general, and I think that this is intuitively clear, our criterion will yield a very 'fine-grained' inventory of events.

We have characterized distinctness of events in terms of their nonidentity, and we have provided the criterion (D5) for identity of events. Let us now return to the counterfactual analysis of causation and see where these notions fit in. To get causal dependence of one event upon another from the counterfactual dependence between them, we must restrict the events involved to those that are distinct. Assuming for the time being that the notion of distictness provided in (D4) and (D5) is adequate for our purposes, we could define causal dependence in the following straightforward way:

(D6) Where *c* and *e* are occurrent events, *e* depends causally on *c* iff:
(1) *c* and *e* are distinct events; and
(2) if *c* had not occurred, then *e* would not have occurred.

Let us turn now from the notion of causal dependence to the notion of causation.

If *e* is causally dependent upon *c*, then we may say that *c* is *a cause* of *e*; however, from the fact that *c* is a cause of *e* we may not conclude that *e* is causally dependent upon *c*. Often, when one event is a cause of another, this is so because there is a causal *chain* of events linking them. Suppose *c* and *e* are the first and the last members of a chain of events, $c, d_1, d_2, \ldots, d_n, e$ such that *c* is a cause of d_2, and d_n is a cause of *e*. Then, we would want to say that *c* is also a cause of *e* (the relation is transitive), but *e* may not be causally dependent upon *c*, since *e* might have occurred even if *c* had not occurred. Consequently, causation between events cannot simply be defined as causal dependence between those events. To get causation, we need the notion of a causal chain. This is easily constructed as follows:

(D7) Where $c, d_1, d_2, \ldots, d_n, e$ is a sequence of occurrent events (but

not necessarily a temporal sequence, and where *c* and *e* may be the only members), this sequence is a *causal chain* iff:
(1) d_1 depends causally on *c*; and
(2) d_2 depends causally on d_1; and . . . .
(*n*) d_n depends causally on d_{n-1}; and
(*n*+1) *e* depends causally on d_n.

Given this notion of a causal chain, we can put forth the following definition of causation:

(D8) Where *c* and *e* are specific events that occurred, *c* is a cause of *e* iff there is a causal chain of occurrent events from *c* to *e*.[10]

In (D8) we have what might be called the core of the counterfactual analysis of '*c* is a cause of *e*'. In another paper[11] I have argued at length that this core analysis needs to be supplemented with further conditions in order to deal with two very important problems. First, there is the problem of guaranteeing the direction of the causal relation. If one event, *c*, is a cause of another, *e*, then we cannot also say that *e* is a cause of *c*. The conditions in (D8) do not guarantee that this is so for all cases; in my expanded analysis I have supplemented (D8) in a way that guarantees this result. Second, there is the problem of causal overdetermination. I distinguish two kinds of overdetermination, namely, pseudo-overdetermination and genuine overdetermination. In cases of the pseudo-overdetermination variety, there are two (or more) events, say *c* and *c'*, such that either of them would have been a cause of some third event *e* if the other had not occurred, but where only one of the two is properly called a cause of *e*. In cases of the genuine variety, there are again two events, *c* and *c'*, such that either of them alone would have been a cause of *e* but where it is appropriate to call both of them a cause of *e*. Cases of genuine overdetermination are particularly troublesome, for in those cases the causal chain requirement in (D8) is not satisfied; that is, the conditions in (D8) do not provide a necessary condition for our causal relation. In my expanded version of (D8), I provide a straight-forward revision which deals with these problems. Since the considerations leading to these revisions, and the revisions themselves, are quite complicated, I shall not present them here. Rather, I shall for simplicity assume that the direction of the causal relation is given in each case to be considered, and I shall assume that none of our examples are situations in which overdetermination is a factor to be dealt with. Even within the framework of these assumptions, however, (D8) is subject to the difficulties announced at the beginning of this paper. I turn now to a consideration of some examples that will illustrate the problem.

II. COMPOUND EVENTS

There are strong analogies to be made between events and physical objects.[12] For example, we say of physical objects that they occupy space-time regions. It is natural to say the same thing about events. We give names to physical objects, and we also give names to events. One very important analogy leads to the first of our troublesome cases for the counterfactual analysis (D8). Many physical objects (indeed, virtually all of those that we are perceptually acquainted with) are compound objects having other physical objects as their components. A brick wall is a physical object, and it has as components each of the bricks and bits of mortar that make it up. Each brick, and each bit of mortar, is in turn a physical object, and each is in turn a compound physical object. Now it is notoriously difficult to say what distinguishes collections of physical objects that are compound objects from other collections that are not compound objects. If you put all of the bricks together in the right way using the mortar, then you have a new physical object, the brick wall. But if you just put those bricks and the mortar out in the yard helter-skelter, then you have a collection of physical objects but no compound object. Exactly similar things can be said about events. Many events (indeed, virtually all of those that we are perceptually acquainted with) are compound events, having other events as their components. As with physical objects, it is very difficult to say what the difference is between a collection of events which is is a compound event and a collection of events which is not a compound event. Suppose, for example, that I type the word *Larry.*[13] This action of mine is a compound event having among its component events the event (action) of my typing the letter *r* twice in succession. Similarly, suppose that my Christmas tree is rigged with a device that flashes the lights on and off every few seconds. Consider a given flashing of the lights on the tree. This is a compound event (but not an action) and it has as its component events the individual flashings of each bulb on the tree. Now suppose that I type the word *Larry* and you quite coincidentally also type the word *Larry*. Then we have two typings of the word, but they do not together constitute a compound event. Or, suppose that my neighbor's Christmas tree is also rigged with a flasher, and on some occasion both trees flash at the same time. Then we have two flashings of trees, but no compound event.

I do not know of any satisfactory way of providing conditions that will distinguish collections of events that form a compound event from collections of events that do not form a compound. One way to approach the problem is to specify a class of what we might call 'event-forming operations' such that, for any set of events, if this class yields an operation that can be performed on the members of the set, then the members form a compound event. Then, we could define the relation '. . . is a component of —' in the following way:

(D9) Event *e* is a component event in the compound event *e*′ iff: there is some event-forming operation (designated by '#') such that, for some set of events a1, a2, . . . , a*n* (possibly having only one member), *e*′ is identical with *e* # a1 # a2 # . . . # a*n*.

I freely admit that, without further specification of the members of the class of event-forming operations, this definition does not advance our understanding of compound events very much! In effect, I am relying upon an intuitive grasp of the notion of a compound event, and will hope that this does not seriously cloud the points that I want to make in this paper.

It is the relation between a compound event and its component events that leads to trouble for our definition (D8). For, some compound events are such that they are counterfactually dependent upon at least some of their component events. Consider this example again:

(e1) *S*'s typing the word *Larry;*
(e2) *S*'s typing the letter *r* twice in succession.

Given (D8), we get the result that the event e2 is a cause of the event e1. First, it is clear that e1 depends counterfactually on e2; if *S* had not typed the letter *r* twice in succession, then *S* would not have typed the word *Larry.* Second, the events e1 and e2 are not identical, according to our criterion (D5). There are surely possible worlds in which *S* types the letter *r* twice in succession without typing the word *Larry.* Given that these events are not identical, it follows that they are distinct, in accordance with our assumed definition of distinctness. Third, in accordance with (D6), we get the result that e1 depends causally on e2; this follows from the two considerations just mentioned. Fourth, in accordance with (D7), there is a causal chain from e2 to e1. In this instance, the causal chain has only the two 'links' in question; there are no intermediate 'links' involved. Finally, from all of these things we get the result that e2 is a cause of e1 in accordance with (D8). Unfortunately, all of these results concerning causal dependence, causal chains, and causation are mistaken. There is *no* causal connection between e2 and e1.

If it is not obvious that there is no causal connection between e2 and e1, the following argument should suffice to make it obvious. As we have noted several times, there is a requirement generally agreed upon that no event is a cause of itself. There is another general principle concerning causation which is, I believe, at least as obvious, namely, that if an event *c* is a cause of another event, *e*, then *c* is a cause of every component of that event *e*. In the case at hand, whatever is a cause of the event of *S*'s typing the word *Larry* (e1) is a cause of each component of that event. Putting these two principles together, if we were to say that e2 is a cause of e1, then we would be committed to saying that e2 is a cause of every component event of e1. However, e2 is itself one of the components of e1, and so, by the second principle, we would get the result that e2 is a cause of itself, which violates the first of our principles.

The claims made in this argument must not be confused with other issues. For example, I have claimed that if an event *c* is a cause of another (compound) event *e*, then *c* is a cause of each of the component events of *e*. This is not the same as saying that if *c* is a cause of some component event *d* of the compound event *e*, then *c* is a cause of the compound event. It may be, for example, that *c* and *d* are both component events of the compound event *e*, and that *c* is a cause of *d*. If so, then even though *c* is a cause of some component event of the compound event *e* (namely, *d*) *c* is not a cause of the whole compound event *e*, for this would imply that *c* is a cause of itself. To take a simple example, consider a row of dominos set up so that when the first falls it is a cause of the second's falling, which is in turn a cause of the third's falling, and so forth. There is then a compound event, *e*, which is the falling of the row of dominos, and it consists of the first domino falling, the second falling, and so on. These component events are related in such a way that they form a causal sequence. Whatever causes the entire compound event is a cause of each falling of a domino, but it does not follow that each event which is a cause of some falling of a domino is a cause of the entire compound event. To suppose this would lead to the result that some falling of a domino is a cause of itself.

One kind of example illustrating the considerations in the previous paragraph has been given some special attention in the literature on action. According to Alvin Goldman, some human actions are *generated* by others even though there is no causal connection between them.[14] The following pair of actions is an example of what Goldman calls *causal* generation:

(e3) *S*'s flipping of the switch;

(e4) *S*'s turning on the light.

Goldman notes, quite correctly, that e3 is not a cause of e4. However, the event e4 is a compound event having the following event, in addition to e3, as one of its component events:

(e5) the light's going on.

Goldman calls this an instance of causal generation because e3 is a cause of the event e5, and it is in virtue of this casual connection that we can ascribe the action e4 to *S*. In the terminology that I am developing here, we can say that e4 is a compound event, having e3 and e5 as component events. We can say that there is a causal connection between the *component* events, but *not* between the component events and the compound event that they form. Just as with the event pair e1 and e2, however, the counterfactual analysis (D8) would yield the result in this causal generation example that e3 is a cause of the compound event e4.

These examples involving compound events and their components indicate, as I have suggested, that we must strengthen our notion of distinctness of events in order to achieve a satisfactory definition of '*c* is a cause of *e*'. It would be sufficient to handle the cases discussed in this section if we revise the definition (D4) of distinctness in the following way:

(D4′) Two events, *c* and *e*, are distinct iff:
(1) *c* and *e* are not identical; and
(2) neither event is a compound event of which the other is a component.

If we revise our definition in this way, then we no longer get the result that e1 depends causally on e2, nor would we get the result that e4 depends causally on e3. Unfortunately, this revision is still not sufficient to avoid *all* of the problems concerning noncausal counterfactual dependencies that face our definition (D8).

III. COMPLEX EVENTS

Consider the following pair of events:

(e6) Socrates' dying;
(e7) Xanthippe's becoming a widow.

Events like the one e7 have been dubbed 'Cambridge changes' by Geach, and are discussed by Kim.[15] Such events present a problem for our definition (D8), even when the supporting definition (D4) is revised along the lines of (D4′). First, let us note that e6 and e7 are not identical events. It is logically possible that Socrates should have died without Xanthippe becoming a widow; just imagine a world, for example, in which they are not married. Similarly, it is logically possible for Xanthippe to have become a widow without Socrates dying; imagine a world, in which Xanthippe was married to someone else, who died. Second, the events e6 and e7 are not related as compound to component. For this relation to hold between them there would, given (D9), have to be some third event *a* (or set of events) and an appropriate even-forming operation '#' such that e7 is identical to e6 # *a*. But, as far as I can see, this is not so. What third event is there such that Xanthippe's becoming a widow is identical with Socrates' dying # this third event? Third, we may note that e7 is clearly dependent counterfactually on e6; if Socrates had not died, then Xanthippe would not have become a widow. These things together suffice to yield the result, given (D8) and the modified definition of distinctness (D4′), that e6 is a cause of e7. But, as in our example involving compound events, this result is incorrect.

There are other event pairs that raise a similar sort of problem for the counterfactual analysis (D8). In the previous section it was noted that Goldman calls the relation between the events

(e3) *S*'s flipping of the switch;
(e4) *S*'s turning on the light,

an instance of "causal generation" of one event (action) by another. We also noted that such cases are plausibly construed as instances of the compound-component event relation. Goldman also characterizes three other kinds of event generation, and each of these can be seen to present a problem for (D8).[16]

First, there is what Goldman calls *conventional* generation. One of his examples involves the events (actions):

(e8) *S*'s extending his arm out of the car window;
(e9) *S*'s signaling to turn.

The event e8 conventionally generates the event e9, according to Goldman, because there is a conventional rule *R* (concerning what counts as a turn signal) and a set of circumstances *C* (approaching a corner, and so on) such that a description of e8 as occurring in *C* in conjunction with the rule *R* entails that e9 occurs. This pair of events is a problem for (D8) since e8 and e9 are not identical, given (D5), nor is e8 a component of e9, given (D9), and yet e9 is counterfactually dependent upon e8. This would give us the unfortunate result that *S*'s extending his arm out of the car window was a cause of his signaling for a turn,

Second, Goldman identifies a class of cases that are instances of what he calls *simple* generation. An example of this kind of generation is the following:

(e10) *S*'s jumping 6 feet 3 inches;
(e11) *S*'s outjumping George.

Simple generation is, according to Goldman, just like conventional generation except that there is no conventional rule *R* involved. Again, this class of cases presents a problem for (D8); scrutiny will reveal that e10 turns out, unhappily, to be a cause of e11.

Third, and finally, some event pairs exemplify what Goldman calls *augmentation* generation. As an example, he suggests

(e12) *S*'s extending his arm;
(e13) *S*'s extending his arm out of the window.

Once again, this event pair will turn out to be a case of causal dependence, contrary to what surely must be everyone's intuition to the contrary.

The distinction between these various kinds of generation is not *very* clearly drawn by Goldman; indeed, he admits that the distinctions may not be all that clear. For my purposes, this does not matter very much. I want to borrow Goldman's general notion of generation, but I shall take it as undefined, and I shall not worry about subdivision into causal, conventional, and so on. Moreover, I shall take generation to be a relation that can hold not only between various pairs of actions, but also between some events which are not actions (such as e6 and e7). Some nonidentical events which are not related by the compound-component relation are related by the generation relation; and these, as in our examples, need to be excluded from the class of causally related events.

I suggest that we consider a class of events which is broader than the class of compound events. I shall call these *complex* events, and I shall speak of complex events as having other events as (or, among) their *constituents*. The class of compound events is a (proper) subclass of the class of complex events.

We may define the relation between a complex event and its constituent events in the following way:

(D10) Event e is a constituent event in the complex event e' iff *either*
(1) e' is a compound event and e is a component event of e'; *or*
(2) e' is generated from e; *or*
(3) both (1) and (2).

All of the examples that we have considered thus far turn out to be cases in which one event is a constituent of another complex event. Even though I have not attempted to specify the conditions under which one event is generated by another, I believe that the classification of these examples in terms of the complex-constituent relation is intuitively correct and satisfying. Moreover, if we think of these event pairs in this way, then there is a natural explanation of our hesitancy to say that these pairs are causally related. The explanation is essentially the same as that given in Section II concerning compound events. Consider the example involving the pair

(e6) Socrates' dying;
(e7) Xanthippe's becoming a widow.

If we take it, as I am suggesting, that e7 is generated by e6, and hence that e6 is a constituent of the complex e7, then we can *argue* that e6 is not a cause of e7 in the following way. Whatever is a cause of a complex event is a cause of every constituent event of that complex event. But, no event is a cause of itself. If e6 were a cause of e7, then since e6 is a constituent of e7, e6 would be a cause of itself. Thus, e6 is not a cause of e7. A similar line of argument can be given in each of the other examples considered.

I suggest, then, that we replace the definition (D4′) of distinctness with the following even broader definition:

(D4″) Two events, c and e, are distinct iff:
(1) c and e are not identical; and
(2) neither event is a complex event of which the other is a constituent.

With (D4″) at hand, the counterfactual analysis of causation (D8) is rendered immune to the troublesome cases raised by Kim and others. Socrates' dying will not turn out to be cause of Xanthippe's becoming a widow since Xanthippe's becoming a widow is a complex event of which Socrates' dying is a constituent. Similarly, since *S*'s signaling to turn is a complex event of which

S's extending his arm out of the car window is a constituent, the latter will not by our account turn out to be a cause of the former.

I have been concerned, in part, to show that the adoption of the broad notion of distinctness embodied in (D4″) provides an intuitively natural way of interpreting the appropriate condition in the counterfactual analysis of causation. As I suggested earlier, the problematic cases discussed in this paper present the same sort of problem for other approaches to the analysis of causation. If we were to develop 'core' versions of a regularity view, or a conditions view, or a manipulability view we would find each of them having a distinctness requirement. If the distinctness requirement were interpreted in the narrow manner of (D4), then all of the examples in this paper would, I am quite certain, turn out to be troublesome for each of these approaches. Each of them could be 'saved', however, by adoption of the broad notion of distinctness that I have suggested. The fact that this notion is both efficacious and 'theory-neutral' is, I believe, strong indication that it deserves serious consideration and further work.

Ohio State University

NOTES

[1] See Jaegwon Kim, 'Noncausal Connections', *Noûs* 8 (1974), 41–52.

[2] This approach to the analysis of causation has only recently begun to attract serious attention. Recent formulations of counterfactual analyses of causation can be found in Aardon Lyon, 'Causality', *British Journal Philosophy of Science* **18** (1967), 1–20; David Lewis, 'Causation', *Journal of Philosophy* **70** (1973), 556–567, reprinted in Ernest Sosa (ed.), *Causation and Conditionals*, Oxford University Press, London, 1975, pp. 180–191; and Marshall Swain, 'A Counterfactual Analysis of Event Causation', *Philosophical Studies* **34** (1978), 1–19. As Lewis points out in his article, there is a hint of a counterfactual analysis in David Hume's *An Enquiry Concerning Human Understanding*, ed. by L. A. Selby-Bigge, rev. by P. H. Nidditch, Oxford University Press, London, 1975, Section 7, 'Of the Idea of Necessary Connexion', pp. 60–79.

[3] The most detailed version of Lewis's theory of counterfactuals is in his book *Counterfactuals*, Harvard University Press, Cambridge, Mass., 1973. See also Lewis's papers 'Completeness and Decidability of Three Logics of Counterfactual Conditionals', *Theoria* **37** (1971), 74–85; and 'Counterfactuals and Comparative Possibility', *Journal of Philosophical Logic* **2** (1973), 418–446. In addition to the works of Lewis, see Robert C. Stalnaker, 'A Theory of Conditionals', *American Philosophical Quarterly Monograph Series* **2** (1968), 98–112; and Robert C. Stalnaker and R. H. Thomason, 'A Semantic Analysis of Conditional Logic', *Theoria* **36** (1970) 23–42. The former paper is reprinted in Sosa, *op. cit.*, pp. 165–179.

[4] For some of the problems, and a competitor see John L. Pollock, 'The "Possible Worlds" Analysis of Counterfactuals', *Philosophical Studies* **29** (1976), 469–476.

[5] Goldman uses this term on p. 9 of his book *Human Action*, Princeton University Press, Princeton, N.J., 1976 (first published 1970, Prentice-Hall, Englewood Cliffs, N.J.). In his discussion, Goldman is concerned with rival theories of *action* identity. I am taking the liberty of extending his term to the problem of event identity in general (actions, after all, are events). Goldman suggests this himself in a footnote on p. 3 of his book.

[6] Throughout this paper I use expressions such as '*S*'s turning on the light' as descriptions of events, each of which is associated with an abbreviated name of that event, such as 'a1'; these names are used to refer to the events in the text.

[7] See Goldman, *op. cit.*, esp. Chap. 1. The 'coarse-grained' view has had its most powerful defense in the writings of Donald Davidson; see, for example, 'Actions, Reasons, and Causes', *Journal of Philosophy* **60** (1963), 685–700; 'The Logical Form of Action Sentences', in N. Rescher (ed.), *The Logic of Decision and Action*, University of Pittsburgh Press, Pittsburgh, 1967, pp. 81–96; and 'The Individuation of Events', in N. Rescher *et al.* (eds.), *Essays in Honor of Carl G. Hempel*, Reidel, Dordrecht, 1970, pp. 216–234.

[8] For a very helpful and precise survey of theories of event-identity, and a sophisticated proposal, see Myles Brand, 'Identity Conditions for Events', *American Philosophical Quarterly* **14** (1977), 329–337. Definition (D5) is in the spirit of the one proposed by Brand, although his involves several complications, the inclusion of which do not seem necessary for my purposes in this paper.

[9] The second of these modal claims is less clear than the first. Is it logically possible for *the* event a2 to have occurred without the event a1 occurring? Obviously, it is possible (logically) for *S* to have alerted the prowler in some *other* way. But then, would that not *necessarily* have been a *different* (i.e., ≠ a2) prowler-alerting? The answer depends upon considerations concerning event constitution that I am not prepared to provide. Fortunately, all we need is the logical possibility of a1 occurring without a2 to get a1 ≠ a2, and I think the logical possibility is relatively uncontroversial. Similar considerations will infect many of the examples to be discussed in this paper.

[10] This definition, as well as the definition (D7) of a causal chain, is essentially the same as the view suggested by David Lewis in 'Causation', *op. cit.*, and by Swain, *op. cit.*

[11] Swain, *op. cit.*

[12] This is emphasized in Brand, *op. cit.*

[13] This example, and several of those to follow, is suggested by Kim in 'Causes and Counterfactuals', *Journal of Philosophy* **70** (1973), 570–572, as a counterexample to Lewis's counterfactual analysis of causation.

[14] See Goldman, *op. cit.*, pp. 22–25.

[15] See Peter Geach, *God and the Soul*, Routledge and Kegan Paul, London, 1969; and Kim, 'Noncausal Connections', *op. cit.*

[16] See Goldman, *op. cit.*, Chap. 2. Goldman intends his classification to apply to *actions*. I am taking the liberty of extending his notions to events in general, although it is not clear that each of his kinds of generation has application outside the sphere of action.

CARL GINET

THE CONDITIONAL ANALYSIS OF FREEDOM

> ... determinism cannot be reconciled with our datum ... that it is sometimes up to me what I am going to do. For if it is ever really up to me whether to do this thing or that, then ... each alternative course of action must be such that I can do it, in the sense that it is then and there within my power to do it. But this is never so, if determinism is true, for on the very formulation of that theory whatever happens at any time is the only thing that can then happen, given all that precedes it. It is simply a logical consequence of this that whatever I do at any time is the only thing I can then do, given the conditions that precede my doing it.
>
> Richard Taylor[1]

I

This is well said. But many philosophers do not see it that way. It is an uncomfortable view to take. It means that a fundamental assumption of our practical life – that we are continually free to determine which of several alternative courses events will take – is not known to be true, since determinism – the thesis that the entire state of the universe at any given time can be deduced from its state at any earlier time and the laws of nature – is not known to be false. It is not surprising that many philosophers believe that this assumption must be compatible with determinism. Many of them have hoped to make a convincing case for this compatibility by giving an analysis of what it means to say that a person could have brought about what in fact that person did not bring about. At least they have hoped to spell out a proposition that will necessarily be the same in truth-value as this one and also be clearly compatible with determinism.

The leading suggestion regarding such a proposition is that it, or a primary component of it, should be a subjunctive conditional of a certain sort: a person *S* could have brought about a certain thing that *S* did not in fact bring about, the suggestion goes, provided that *S* would have done so if *S*'s will had

Peter van Inwagen (ed.), Time and Cause, 171–186.
Copyright © 1980 *by D. Reidel Publishing Company.*

been in the appropriate direction. It is a plausible idea that *S* was free to do something that *S* did not do if the circumstances lying outside *S*'s actual will and the factors motivating it would not have frustrated *S*'s will had it been towards doing that thing.

Concerning this suggestion Taylor has remarked that, if determinism is true, then

> I could not have decided, willed, chosen, or desired otherwise than I in fact did . . . we will then want to know whether the causes of those inner states were within my control; and so on, *ad infinitum*. We are, at each step, permitted to say "could have been otherwise" only in a provisional sense . . . but must retract it and replace it with "could not have been otherwise" as soon as we discover, as we must at each step, that whatever would have to have been different could not have been different.[2]

To deal with this sort of objection a sophisticated compatibilist will include in the proposed analysis, in addition to the sort of subjunctive conditional mentioned above, another clause the import of which is intended to be that *S* could have had a different will. And the compatibilist will see to it that this added clause, besides being compatible with determinism, is not a subjunctive conditional and thus will not permit the challenge, 'But could *S* have made *that* proviso the case?'[3] By adding such a nonconditional necessary condition the compatibilist will also avoid most of the other objections to conditional analyses of freedom that have been put forward.[4]

The compatibilist will, however, still not have an acceptable analysis of the conditions necessary and sufficient for its having been in a person's power to make the world other than it was. No matter what sort of clause is added, as long as it remains compatible with determinism, the proposed analysis will have clearly unacceptable consequences. This is the conclusion that I will try to make credible. In addition, I will make a suggestion as to what an acceptable conditional analysis of freedom should look like.

II

Our analysandum, in its most general form, can be expressed in the following way.

It was in *S*'s power at *t* to make it the case that *p*.

S had freedom of the sort we are concerned with only if some propositions of this form are true when the embedded proposition *p* is false. The proposition *p* may express the occurrence of an event (or events) or state of affairs (or states of affairs) of virtually any sort (or sorts). We attribute to people, not only the freedom to determine which of several alternative actions they will perform, but also the freedom to determine which of several alternative events or states of affairs of other sorts will obtain. Such freedom implies the power to make the case what is not the case.

To make something the case is to contribute at least part of what is needed for it to be the case. More precisely, S made it the case that p just in case p is true and there was some action of S's, S's V-ing, such that either p entails that S V-ed or there is some proposition q such that S's V-ing caused it to be the case that q and p entails q.

The power to make a certain thing occur at a certain time, though possessed at one time, can later be lost as matters change and the time of the thing in question draws nearer. For example, it might be that S, sitting in her office, had it in her power thirty minutes ago to be on the squash court at 4.00 P.M. but now, at 3.55 P.M., no longer has that in her power. Since the relation between the time at which S had the power to make a certain state of affairs obtain (or event occur) and the time of that state of affairs (event) is important, we will do well to make reference to the time of the state of affairs (event) explicit in our analysandum. It will also be a good idea to be fairly precise in what we mean by the occurrence at a time of a state of affairs or event. Let us use 't', with or without subscript, to denote an interval (which may be vaguely delimited) located at a particular place in time. Let us use '$F(t)$' to abbreviate the sentence form 'Such-and-such a thing(or things) was(were) in such-and-such a state, or undergoing such-and-such a change, from the beginning to the end of t'.

This represents one form of proposition that may replace p in our general analysandum. It also makes sense to say that it was in S's power at t to make the case a certain *conjunction* of states of affairs or events, which may (but need not) be temporally scattered. Thus a more inclusive specification of the sort of proposition that may replace p is given by the phrase 'a conjunction of one or more propositions each of the form $F(t)$'. When there is more than one such conjunct it will be convenient to have them conjoined in temporal order. So let us represent this more inclusive form of proposition in the following fashion:

$$F_1(t_1) \& \ldots \& F_n(t_n),$$

where $n \geqslant 1$ and $t_1, \ldots, t_n$ are so ordered that for each t_i other than t_n, t_{i+1} does not begin earlier than t_i (I will use '$t_i \leqslant t_{i+1}$ to express this relation). Let us limit the task under discussion to that of providing a necessary and sufficient condition for the truth of our analysandum in just those cases where p is some proposition of the form just specified. (Other forms of replacements for p could be constructed by using quantification and truth-functional connectives other than conjunction, but we need not consider them here.)

We can think of the prefix 'It was in S's power at t to make it the case that . . .' as expressing an operation on the proposition expressed by the sentence to which it is prefixed. We can think of it as expressing a certain kind of possibility for that proposition, a kind that is relative to a particular person and time. Let us abbreviate it with the following:

$$\langle\text{P}\rangle^{S}_{t}.$$

Let us say that '$\langle\text{P}\rangle^{S}_{t}\, p$' expresses the *power possibility* of the proposition p relative to the time t and the person S.

We can, if we like, introduce the notion of the *power necessity* of a proposition, relative to a particular person and time, defining it in terms of power possibility:

$$\boxed{\text{P}}^{S}_{t}\, p =_{\text{df.}} p \;\&\; {\sim}\langle\text{P}\rangle^{S}_{t}{\sim}p.$$

And we can define an unrelativized notion of power necessity in terms of the relative notion:

$$\boxed{\text{P}}\, p =_{\text{df.}} \wedge S \wedge t\; \boxed{\text{P}}^{S}_{t}\, p.$$

Using some of our new notation, we can represent the analysandum that we are now concerned with as follows:

$$[1] \qquad \langle\text{P}\rangle^{S}_{t}\, [F_1(t_1) \;\&\; \ldots \;\&\; F_n(t_n)].$$

This is to be read: it was in S's power at t to make it the case that F_1 occurred at $t_1, \ldots,$ and F_n occurred at t_n.

III

A philosopher seeking a compatibilist analysis of the truth-conditions for [1] can, perhaps, do no better than to begin by working out an analysis for the simpler one-conjunct case,

$$[2] \qquad \langle\text{P}\rangle^{S}_{t}\, F(t_1),$$

and then try to adapt that analysis to fit the more general case, [1].

It is clear that no proposition of the form of [2] can be true unless *S* exists at *t*. *S* possesses no powers at times when *S* does not exist. Further, it is clear that [2] can be true only if t is not wholly later than t_1. Once a time is past it is no longer in one's power to make something the case at that time. Finally we can agree that [2] is true only if either *S* did make it the case that $F(t_1)$ by something *S* did between the beginning of t and the end of t_1 or *S* would have done so if *S*'s will had been suitably different from what it was between those times.

The term 'will' here refers to whatever suffices to initiate an intentional action or attempt to act that involves exertion of force by the body. If one is persuaded (as I am) that this initial component must be a *volition*, a mental act of willing to exert force with one's body in a certain way, then one will find it appropriate to replace the vague 'if *S*'s will had been suitably different' with 'if *S* had performed a volition (or volitions) of appropriate type(s)'. Many philosophers, however, are not persuaded that intentional actions or attempts to act must begin with volitions. Since none of the difficulties for the compatibilist program of analysis that I shall raise depend on taking one position rather than the other on this question, some readers will be spared unnecessary annoyance if we use a formulation of the subjunctive conditional (at the end of the preceding paragraph) that does not presuppose any particular answer to it. So instead of volitions I propose to refer to *causally basic actions* and to define the latter notion as follows. An action (e.g., my signing a check) is *not* causally basic just in case it consists of (i) an event *e* (e.g., there coming to be certain marks on the check) that is not itself an action, and (ii) another action of the same agent (e.g., my moving a pen in a certain way) that does not have *e* as a part but does contribute to causing *e*.[5] Philosophers who do not accept that an action that is causally basic in this sense must be a volition can still agree that this notion offers a suitable way to make more specific the antecedent of the subjunctive conditional that is necessary for the truth of [2] when $F(t_1)$ is false.

A careful formulation of this subjunctive conditional must allow for the possibility that making $F(t_1)$ true would have required the performance by *S* of more than one causally basic action. I would have made it the case that the broken step in front of my house is now repaired had I performed a fairly large number of different causally basic actions at different times, but there is (alas) no single causally basic action such that if I had merely performed it at a certain past time then I would have made it the case that that step is now repaired. More accurately, there is no type of causally basic action such that if I had merely performed a single action of that type then by doing so I would have made it the case that that step is now repaired. So I propose the following as the form of our subjunctive conditional:

[3] There are times $t_{1.1} \leqslant t_{1.2} \leqslant \ldots \leqslant t_{1.m}$, between the beginning of t and the end of t_1, and causally basic action types $A_{1.1}, \ldots, A_{1.m}$, such that: if S had performed $A_{1.1}$ at $t_{1.1}, \ldots$, and $A_{1.m}$ at $t_{1.m}$, then by doing so S would have made it the case that $F(t_1)$.

([3] should be interpreted in such a way that it is true if the antecedent and consequent of the embedded conditional were in fact both made true by one or more times and action types meeting the conditions laid down.)

IV

Incompatibilists can agree with compatibilists that [3] is necessary for [2]. Disagreement comes over what more is necessary. A small part of the answer is, however, unlikely to be disputed. The antecedent of the subjunctive conditional in [3] must be logically possible and it must also be logically compatible with the laws of nature, i.e., naturally possible. Otherwise, if we allow [3] as it stands to be sufficient for [2], we allow it to be in S's power to perform actions that are logically or naturally impossible. But, of course, it can be in no one's power to make the case a logically impossible state of affairs. And it can be in no natural person's power to make the case a naturally impossible state of affairs. Part of the notion of natural necessity that attaches to laws of nature and to their logical consequences is the notion of unrelativized power necessity defined above. A proposition is naturally necessary only if it is never in any natural being's power to make true its contradictory. At any rate, the determinism that I wish to say is incompatible with our ever being free to make things other than they are is the doctrine that every state of the universe develops from antecedent states in accordance with universal laws of nature that are necessary at least in the sense that nothing in nature has the power to falsify them. So we can agree that the following elaboration of [3] is necessary for [2].

[4] There are times $t_{1.1} \leqslant \ldots \leqslant t_{1.m}$, between the beginning of t and the end of t_1, and causally basic action types $A_{1.1}, \ldots, A_{1.m}$, such that:

(i) if S had performed $A_{1.1}$ at $t_{1.1}, \ldots$, and $A_{1.m}$ at $t_{1.m}$, then by doing so S would have made it the case that $F(t_1)$, and

(ii) it is logically and naturally possible that *S* performed all of the actions mentioned in the antecedent of (i).

What more is necessary? Well, these conditions do not guarantee that it was ever in *S*'s power to perform any of the sequence of causally basic actions mentioned in (i). For any such sequence to have been in *S*'s power, it must be, or be part of, a sequence whose first member was in *S*'s power up to the time when it would have occurred. For, if the performance of a causally basic action *A* at *t* was in *S*'s power up to t^-, earlier than *t*, but ceased to be so after t^-, then this could only be because *S* failed to perform at t^- some causally basic action A' that it was in *S*'s power up to t^- to perform then and the performance of which would have preserved, past t^-, *S*'s power to perform *A* at *t*. From this we can see that what is required for the truth of [2] is a sequence of (one or more) causally basic actions, $A_{1.1}$ at $t_{1.1}, \ldots, A_{1.m}$ at $t_{1.m}$, such that: $\langle\mathrm{P}\rangle^S_{t_{1.1}} A^S_{1.1}(t_{1.1})$, if $A^S_{1.1}(t_{1.1})$ then $\langle\mathrm{P}\rangle^S_{t_{1.1}} A^S_{1.2}(t_{1.2})$, $\ldots$, and if $A^S_{1.m-1}(t_{1.m-1})$ then $\langle\mathrm{P}\rangle^S_{t_{1.m}} A^S_{1.m}(t_{1.m})$, where '$A^S(t)$' abbreviates '*S* performed *A* at *t*'. We cannot, of course, include *this* condition in the analysis without rendering it circular. We must consider what is necessary and sufficient for the truth of any of its atomic components containing '$\langle\mathrm{P}\rangle$', any proposition that it was in *S*'s power up to a particular time *t* to perform a certain causally basic action at *t*.

I am prepared to argue (and by the end of this paper will have argued) that it is necessary for the truth of such a proposition that no facts as to what occurred up to *t* naturally necessitate that *S* did *not* perform *A* at *t*. The compatibilist thinks that we need not go so far, that what is necessary is the absence, not of any sort of fact that naturally necessitates *S*'s not performing *A* at *t*, but just the absence of certain restricted sorts of such facts. *S*'s not performing *A* at *t*, the compatibilist will say, must not have been naturally necessitated by antecedent facts that were both *external* to *S*'s motivational self and *too close* causally *to t*. If determinism is true then, if one goes back far enough one will find antecedent facts quite external to *S* that causally necessitated *S*'s not performing *A* at *t*. But the compatibilist's idea is that this does not conflict with its having been in *S*'s power up to *t* to perform *A* at *t* if those antecedents necessitated *S*'s not performing *A* at *t through* events and states that were among those desires, beliefs, intentions, and the like with which *S* 'identified' himself or herself. Thus the compatibilist will want to formulate a necessary and sufficient condition for the truth of $\langle\mathrm{P}\rangle^S_t A^S(t)$ that will be satisfied in cases where *S*'s not performing *A* at *t* but performing some other causally basic action instead is a matter of the actual action's arising in the normal way (whatever that is) from S's immediately antecedent desires, beliefs, and the like. But the compatibilist will want this condition also to be such that it will *not* be satisfied in cases where external circumstances, bodily conditions, or even desires and fears of a special compulsive character, figure in the explanation of S's not performing A at t in such a way that it is intuitively clear that for some period before t it was not in S's power to perform A at t. Since it does not matter for the difficulties I am going to raise what informative, and at least superficially plausible, version of such a condition is used, let us allow the compatibilist to put it vaguely, but simply, as follows:

> S's not performing A at t was not naturally or otherwise necessitated by conditions up to t that were external to S's motivational self and too near causally to t.

(I include 'or otherwise' here in case it turns out that, for instance, the compulsion in compulsive desires and fears is not to be explicated in terms of natural necessitation.) We can take it on faith that the cash value of 'otherwise', 'external to S's motivational self', and 'too near causally to t' could be spelled out in plausible fashion. It will be convenient to have an abbreviation for this condition. I shall use '$\mathrm{O}[A^S(t)]$'.

We can agree, then, that the following elaboration of [4] is necessary for the truth of [2].

[5] There are times $t_{1.1} \leqslant \ldots \leqslant t_{1.m}$, between the beginning of t and the end of t_1, and causally basic action types $A_{1.1}, \ldots, A_{1.m}$, such that:
(i) if S had performed $A_{1.1}$ at $t_{1.1}, \ldots$, and $A_{1.m}$ at $t_{1.m}$ then by doing so S would have made it the case that $F(t_1)$, and
(ii) $\mathrm{O}[A^S_{1.1}(t_{1.1})]$, if $A^S_{1.1}(t_{1.1})$ then $\mathrm{O}[A^S_{1.2}(t_{1.2})], \ldots$, and if $A^S_{1.m-1}(t_{1.m-1})$ then $\mathrm{O}[A^S_{1.m}(t_{1.m})]$.

[5] (ii) entails [4] (ii) on the natural stipulation that a proposition of the form '$\mathrm{O}[A^S(t)]$' is to be counted as false if $A^S(t)$ is logically or naturally impossible. It will be convenient to have an abbreviation of [5]. Since [5] expresses a certain relation among S, the times t and t_1, and the proposition $F(t_1)$, let us use the following: $\mathrm{R}[S,t,t_1,F(t_1)]$.

Now the compatibilist might suggest that, if one conjoins [5] with the stipulation that S existed at t, one gets a condition that is sufficient as well as necessary for the truth of [2], namely,

[6] S existed at t and $\mathrm{R}[S,t,t_1,F(t_1)]$.

V

This claim looks plausible. Now the compatibilist's task is to show how this result can be used to construct an analysis for the more general case, [1], where the embedded proposition can have any number of conjuncts (greater than zero). I am not aware of any attempt by a compatibilist to tackle this problem. It is where the insuperable difficulties arise. It will not do, of course, to say that

$$\langle \mathrm{P} \rangle_t^S [F_1(t_1) \,\&\, \ldots \,\&\, F_n(t_n)] \text{ iff } \langle \mathrm{P} \rangle_t^S F_1(t_1) \,\&\, \ldots \,\&\, \langle \mathrm{P} \rangle_t^S F_n(t_n).$$

It must be possible that it was in S's power at some time to determine *whether or not* a certain state of affairs would occur at a certain time. That is, it must be possible that, for some S, t, and proposition $F(t_1)$,

$$\langle \mathrm{P} \rangle_t^S F(t_1) \,\&\, \langle \mathrm{P} \rangle_t^S {\sim}F(t_1).$$

But, of course, it is not possible that

$$\langle \mathrm{P} \rangle_t^S [F(t_1) \,\&\, {\sim}F(t_1)].$$

Perhaps the simplest move that is not so obviously wrong that it would never be seriously considered is to replace '$F(t_1)$' in [6] with '$F_1(t_1)$ & ... & $F_n(t_n)$'. This would give us the following candidate for a condition that is necessary and sufficient for the truth of [1]:

[7] S existed at t and $\mathrm{R}[S,t,t_1,F_1(t_1) \,\&\, \ldots \,\&\, F_n(t_n)]$.

This candidate must, however, be rejected. In some cases it gives wrong results. Consider a two-conjunct case,

[8] $\langle \mathrm{P} \rangle_t^S [F_1(t_1) \,\&\, F_2(t_2)]$,

and suppose the following: $F_1(t_1)$ is true, $F_2(t_2)$ is false but $\langle \mathrm{P} \rangle_t^S F_2(t_2)$ is true, and t_1 is wholly earlier than t. It is consistent with these suppositions that [8] is true. Yet [8] could not be true if the appropriate instance of [7] were necessary for [8]. The appropriate instance of [7] is not satisfied by the case supposed because there are no times between the beginning of t and the end of t_1, owing to the fact that t_1 is wholly earlier than t (see [5] above, p. 178).

One might be tempted to deny that the truth of [8], or of any instance of [1], is compatible with the supposition that t_1 is wholly earlier than t, and to insist that if it is in *S*'s power at t to make it the case that p then p cannot entail any proposition as to what happened before t. And one might think to argue for this from the assumption that relative power possibility, like logical possibility, is preserved by deduction, that if relative power possibility attaches to a given proposition then it must attach to any logical consequence of that proposition. But this assumption is false. Consider the proposition:

[9] Today is the 87th consecutive day that *S* has run a mile.

It could be that it is in *S*'s power now to make [9] the case. That is, it is logically possible that

[10] $\langle \mathrm{P} \rangle^{S}_{\mathrm{now}}$ today is the 87th consecutive day that *S* has run a mile.

Proposition [10] might be colloquially expressed by saying, '*S* could make this the 87th straight day she has run a mile'. [9], however, has among its logical consequences the proposition that

[11] *S* began to exist at least 86 days before today.

In no possible world where [10] is true, indeed in no possible world at all, is it true that it is in *S*'s power now to make [11] true. Thus it is false that if relative power possibility attaches to a proposition then it attaches to any logical consequence of it.

So the claim that [8] or [1] can be true only if t_1 is not wholly earlier than t cannot be defended in that way. And it is easy to see that it cannot be defended at all, for the logical possibility of [10] also offers the basis for a direct counterexample to this claim. Proposition [9] is equivalent to a conjunction of the form $F_1(t_1)$ & . . . & $F_n(t_n)$ having as its first conjunct the proposition that

on the 86th day before today *S* ran a mile.

Thus [10] is equivalent to an instance of the form [1] where t_1 is wholly earlier than t.

VI

How might [7] be revised to avoid the counterexample to the claim that it is necessary for [1]? Probably the most obvious suggestion is the following: in the case where $F_1(t_1)$ & . . . & $F_n(t_n)$ contains both true and false conjuncts, only the false conjuncts need to be treated along the lines of the second conjunct of [7]; and the times of the causally basic actions on S's part that would have made all of those false conjuncts true should be required merely to fall between t and the latest time in those false conjuncts. In other words, the suggestion would be that, if $F_1(t_1)$ & . . . & $F_n(t_n)$ contains some false conjuncts and $F_j(t_j)$, . . . , $F_k(t_k)$ are all of its false conjuncts, then [1] is true if and only if S existed at t and $R[S,t,t_k,F_j(t_j)$ & . . . & $F_k(t_k)]$.

What about the other case, where $F_1(t_1)$ & . . . & $F_n(t_n)$ contains no false conjuncts? Obviously [1] is not always true in that case; but it seems plausible to suppose that in that case [1] will be true if and only if $F_1(t_1)$ & . . . & $F_n(t_n)$ contains at least one conjunct $F_i(t_i)$ such that $\langle P \rangle_t^S F_i(t_i)$. So it might be suggested, plausibly, that, if $F_1(t_1)$ & . . . & $F_n(t_n)$ is true, then [1] is true if and only if S existed at t and $F_1(t_1)$ & . . . & $F_n(t_n)$ contains at least one conjunct $F_i(t_i)$ such that $R[S,t,t_i,F_i(t_i)]$ – remember that we interpret the subjunctive conditional in the proposition that $R[S,t,t_i,F_i(t_i)]$ in such a way that that proposition is true if S did make it the case that $F_i(t_i)$ by something S did between t and t_i.

If we put these suggestions together, we get the following candidate for a condition that is necessary and sufficient for the truth of [1]:

[12] S existed at t; if $F_1(t_1)$ & . . . & $F_n(t_n)$ contains some false conjuncts and $F_j(t_j)$, . . . , $F_k(t_k)$ are all of the false conjuncts it contains, then $R[S,t,t_k,F_j(t_j)$ & . . . & $F_k(t_k)]$; and if $F_1(t_1)$ & . . . & $F_n(t_n)$ contains no false conjuncts, then it contains at least one conjunct $F_i(t_i)$ such that $R[S,t,t_i,F_i(t_i)]$.

But this candidate will not do either. The supposition that [12] is sufficient for [1] has the disastrous consequence that it can be in S's power to make true a logically impossible proposition. Suppose the following: $F_1(t_1)$ & $F_2(t_1)$ is logically impossible; S made it the case that $F_1(t_1)$; t is wholly earlier than t_1. Consistently with these suppositions, we can also suppose that, for some time $t_{1.1}$, not ending earlier than t or later than t_1, and some causally basic action type $A_{1.1}$, (i) if S had performed $A_{1.1}$ at $t_{1.1}$ then by doing so S would have made it the case that $F_2(t_1)$, and (ii) S's not performing $A_{1.1}$ at $t_{1.1}$ was not naturally or otherwise necessitated by conditions before $t_{1.1}$ that were external to S and too near causally to $t_{1.1}$. For example, let $F_1(t_1)$ be *S's head now faces to S's right*, $F_2(t_1)$ be *S's head now faces to S's left*, t be 3 seconds before now, and *S performed* $A_{1.1}$ *at* $t_{1.1}$ be *During the interval from 3 seconds ago until now S turned S's head from facing forward to facing to S's left*. If [12] were sufficient for [1] then it would follow from our last supposition that $\langle \mathrm{P} \rangle^{S}_{t} F_2(t_1)$; and, since $F_2(t_1)$ is the only conjunct in $F_1(t_1)$ & $F_2(t_1)$ that is false; it would follow further that $\langle \mathrm{P} \rangle^{S}_{t} [F_1(t_1)$ & $F_2(t_1)]$.

VII

There is a simple way to amend [12] to avoid the consequence just demonstrated. Add the stipulation that it be logically and naturally possible that $F_1(t_1)$ & . . . & $F_n(t_n)$. That this is required for the truth of [1] is obvious as soon as one thinks of it. Indeed, it is obvious that something stronger must be required, namely, that it be logically and naturally possible that *S made it the case* that $F_1(t_1)$ & . . . & $F_n(t_n)$ *by something S did* at t or later. So let us add this stronger stipulation to [12]. Do we now have a condition that is necessary and sufficient for the truth of [1]? I am afraid not. There remains the following difficulty. If our revised [12] were sufficient for [1] then the inference rule of *modus ponens* for relative power necessity would not be valid. This rule is the following:

$$\boxed{\mathrm{P}}^{S}_{t}(p \supset q) \;.\; \boxed{\mathrm{P}}^{S}_{t} p \therefore \boxed{\mathrm{P}}^{S}_{t} q.$$

This is equivalent to

$$(p \supset q) \,\&\, {\sim}\langle \mathrm{P} \rangle^{S}_{t}(p \,\&\, {\sim}q) \;.\; p \,\&\, {\sim}\langle \mathrm{P} \rangle^{S}_{t} {\sim}p \therefore q \,\&\, {\sim}\langle \mathrm{P} \rangle^{S}_{t} {\sim}q.$$

Surely this is a valid form of inference, on any reasonable understanding of 'It was in S's power at t to make it the case that' (or 'At t it was open to S to make it the case that' or 'At t, S could have made it the case that'). Suppose, for example, that (first premise) if (p) it rained this afternoon then (q) the fresh paint on the house is ruined, and it was not in S's power this morning (t) to make it the case that it would rain this afternoon but the paint would not be ruined. Suppose also that (second premise) it did rain this afternoon and it was not in S's power this morning to make it the case that it would not rain this afternoon. Surely it follows that (conclusion) the paint is ruined and it was not in S's power this morning to make it the case that the paint would not be ruined.

If, however, the revised [12] were sufficient for [1], and if determinism were true, then there would be an instance of this form of inference with true premises and false conclusion. Let t_2 be a time later than t at which S performed a normal action of causally basic type A. That is, there is some contrary causally basic action type A^* such that S's performing A^* instead of A at t_2 was logically and naturally possible and S's not performing A^* at t_2 was not naturally or otherwise necessitated by conditions before t_2 that were external to S and too near causally. Let t_1 be a time wholly earlier than t. If determinism is true then there was a state of affairs obtaining at t_1, $F(t_1)$, such that it was not naturally possible both that $F(t_1)$ and that S performed A^* at t_2. The following two propositions are clearly true in this case.

[13] $\sim\langle\mathrm{P}\rangle^S_t \sim F(t_1)$

[14] $\sim\langle\mathrm{P}\rangle^S_t [F(t_1)\ \&\ S \text{ performed } A^* \text{ at } t_2]$

Proposition [13] is true because t is wholly later than t_1 and once a time is past it is no longer in anyone's power to make something occur at that time. Proposition [14] is true because '$F(t_1)$ & S performed A^* at t_2' is naturally impossible. The truth of [13] and [14] is entailed by the claim that the revised [12] is *necessary* for [1] and, as far as I can see, there is nothing wrong with this claim. Given that [13] and [14] are true, it is obvious that the following are true.

[15] $F(t_1)\ \&\ \sim\langle\mathrm{P}\rangle^S_t \sim F(t_1)$

[16] $[F(t_1) \supset \sim S \text{ performed } A^* \text{ at } t_2]\ \&\ \sim\langle\mathrm{P}\rangle^S_t [F(t_1)\ \&\ S \text{ performed } A^* \text{ at } t_2]$

By *modus ponens* for relative power necessity (and double negation) [15] and [16] yield the following:

[17] $\sim S \text{ performed } A^* \text{ at } t_2\ \&\ \sim\langle\mathrm{P}\rangle^S_t\ S \text{ performed } A^* \text{ at } t_2.$

But if the revised [12] were *sufficient* for [1] then there would be true in the case hypothesized something that contradicts [17], namely, that

$$\langle\mathrm{P}\rangle^S_t\ S \text{ performed } A^* \text{ at } t_2.$$

VIII

The importance of the validity of *modus ponens* for relative power necessity goes considerably beyond showing that our revised [12], although necessary, cannot be sufficient for the truth of [1]. It shows, in fact, that no condition that *is* sufficient for [1] could be satisfied in a world where $F_1(t_1)\ \& \ldots \&\ F_n(t_n)$ is false and determinism is true. That is to say, suppose we are given a truth of the form

[18] For any $S,t,F_1(t_1), \ldots, F_n(t_n)$, where $n \geqslant 1$ and $t_1 \leqslant \ldots \leqslant t_n$: if $C[S,t,F_1(t_1) \,\&\, \ldots \,\&\, F_n(t_n)]$ then it follows that $\langle\mathrm{P}\rangle^{S}_{t}[F_1(t_1) \,\&\, \ldots \,\&\, F_n(t_n)]$,

and we are also given an instantiation of the form $F_1(t_1) \,\&\, \ldots \,\&\, F_n(t_n)$ that is false in some world W where determinism is true. Then, for the same instantiation of that form, the proposition that $C[S,t,F_1(t_1) \,\&\, \ldots \,\&\, F_n(t_n)]$, which according to [18] is sufficient for [1], must be false in W if *modus ponens* for relative power necessity is valid.

To see this, suppose the contradictory. Suppose that the proposition that $C[S,t,F_1(t_1) \,\&\, \ldots \,\&\, F_n(t_n)]$ is true in W. Now, since determinism holds in W, there is true in W some proposition $F_0(t_0)$ such that t_0 is wholly earlier than t and $F_0(t_0) \,\&\, F_1(t_1) \,\&\, \ldots \,\&\, F_n(t_n)$ is naturally impossible in W. If [18] is true, then $C(S,t,\sim F_0(t_0)$ must be false in W; because $\langle\mathrm{P}\rangle^{S}_{t}\sim F_0(t_0)$ must be false in W: t_0 is wholly earlier than t and once a time is past it can no longer be in anyone's power to make the case at that time what was not in fact the case at that time. And if [18] is true, then $C[S,t,F_0(t_0) \,\&\, F_1(t_1) \,\&\, \ldots \,\&\, F_n(t_n)]$ must be false in W; because $\langle\mathrm{P}\rangle^{S}_{t}[F_0(t_0) \,\&\, F_1(t_1) \,\&\, \ldots \,\&\, F_n(t_n)]$ must be false in W: in no possible world can a natural being have the power to make the case what is naturally impossible in that world. Therefore, if [18] is true, $C[S,t,F_1(t_1) \,\&\, \ldots \,\&\, F_n(t_n)]$ is true in W, and other things are as hypothesized, then $\langle\mathrm{P}\rangle^{S}_{t}\sim F_0(t_0)$ and $\langle\mathrm{P}\rangle^{S}_{t}[F_0(t_0) \,\&\, F_1(t_1) \,\&\, \ldots \,\&\, F_n(t_n)]$ are false in W, while $F_0(t_0)$, $\sim[F_1(t_1) \,\&\, \ldots \,\&\, F_n(t_n)]$, and $\langle\mathrm{P}\rangle^{S}_{t}[F_1(t_1) \,\&\, \ldots \,\&\, F_n(t_n)]$ are true in W. But this assignment of truth-values in W entails that a certain instance of *modus ponens* for relative power necessity has true premises and false conclusion in W, namely the following instance:

[19] $\boxed{\mathrm{P}}^{S}_{t}[F_0(t_0) \supset \sim(F_1(t_1) \,\&\, \ldots \,\&\, F_n(t_n))] \;.\; \boxed{\mathrm{P}}^{S}_{t}F_0(t_0)$
$\therefore\; \boxed{\mathrm{P}}^{S}_{t}\sim[F_1(t_1) \,\&\, \ldots \,\&\, F_n(t_n)]$.

Therefore, if *modus ponens* for relative power necessity is valid and [19] is true, then $C[S,t,F_1(t_1) \,\&\, \ldots \,\&\, F_n(t_n)]$ must be false in any world where $F_1(t_1) \,\&\, \ldots \,\&\, F_n(t_n)$ is false but determinism is true; and this holds no matter what condition C may be supposed to be. This conclusion is, of course, the thesis of incompatibilism.[6]

IX

The exercise of working out our revised [12] was not in vain, however. As I said, there is nothing wrong, as far as I can see, with the suggestion that our revised [12] is necessary for the truth of [1]. So the task of finding an analysis of the truth-conditions for [1] is partly accomplished. What remains is to determine what we can add to our revised [12] that will make it sufficient while keeping it necessary.

At the moment I can see nothing wrong with the following answer to this: we can replace the condition $O[A^S(t)]$, used in the definition of the relation R, given in [5], with something stronger, namely, the following:

> *S*'s not performing *A* at *t* was not naturally or otherwise necessitated by *any* conditions up to *t*.

Let this be abbreviated by '$O^*[A^S(t)]$'. Call the relation defined by the result of replacing O with O* in [5], R*. Then the suggestion I am now venturing can be expressed as follows: necessarily, [1] is true if and only if:

> *S* existed at *t*; it is logically and naturally possible that *S* made it the case that $F_1(t_1)$ & . . . & $F_n(t_n)$ by something *S* did at *t* or later; if $F_1(t_1)$ & . . . & $F_n(t_n)$ contains some false conjuncts and $F_j(t_j)$, . . . , $F_k(t_k)$ are all of the false conjuncts it contains, then $R^*[S,t,t_k,F_j(t_j)$ & . . . & $F_k(t_k)]$; and if $F_1(t_1)$ & . . . & $F_n(t_n)$ contains no false conjuncts, then it contains at least one conjunct $F_i(t_i)$ such that $R^*[S,t,t_i,F_i(t_i)]$.[7, 8]

Cornell University

NOTES

[1] Richard Taylor, *Metaphysics*, Prentice-Hall, Englewood Cliffs, N.J., 1963, p. 49.

[2] *Ibid.*, p. 44.

[3] See, for example, Wilfrid Sellars, 'Fatalism and Determinism', in Keith Lehrer (ed.), *Freedom and Determinism*, Random House, New York, 1966, pp. 141–174, esp. pp. 171–174.

[4] In papers such as those of J. L. Austin, 'Ifs and Cans', *Proceedings of the British Academy* **42** (1956), 107–132; Lehrer, 'Ifs, Cans, and Causes', *Analysis* **20** (1960), 122–124; R. Chisholm, 'J. L. Austin's Philosophical Papers', *Mind* **73** (1964), 20–25; and Lehrer, 'An Empirical Disproof of Determinism?', in Lehrer (ed.), *op. cit.*, pp. 175–202.

[5] This definition of causally basic action is essentially the same as that proposed by

H. McCann in 'Volition and Basic Action', *Philosophical Review* **83** (1974), 451–473, where it is also argued that causally basic actions in this sense are always volitions.

[6] The argument of this section resembles those in P. van Inwagen, 'The Incompatibility of Free Will and Determinism', *Philosophical Studies* **27** (1975), 185–199; and C. Ginet, 'Might We Have No Choice?', in Lehrer (ed.), *op. cit.*, pp. 87–104.

[7] Even if this suggestion is right as far as it goes, it leaves unfinished the job of giving the informative analysis of the truth-conditions for [1] that we want. Beyond alluding to compulsive desires and fears, I have said nothing to explain what relations other than natural necessitation may be referred to in the definition of O* by the phrase 'naturally or otherwise necessitated'. I think that there are various cases in which other relations are the basis of intuitively compelling judgments that the subject was rendered powerless to perform a certain action.

[8] I am grateful to Sydney Shoemaker for helpful comments on an earlier version. I wish to thank the National Endowment for the Humanities for fellowship support during part of the time that I was working on this paper.

KEITH LEHRER

PREFERENCES, CONDITIONALS AND FREEDOM*

Richard Taylor directed my dissertation on the subject of freedom and determinism, and we amiably and constructively disagreed about the subject then. I was a compatibilist, he was not. I still am; he is not. So be it. I cannot, however, resist another effort. In this paper, therefore, I wish to provide a condition which is sufficient for saying that a person could have done otherwise and which is compatible with the truth of determinism.

I. THE TRADITIONAL ANALYSIS

The attempt to analyse 'could have' statements in terms of conditionals has usually been in terms of some such conditional as '*S* would have if *S* had chosen to' or '*S* would have if *S* had tried to' or '*S* would have if *S* had wanted to' and so forth. As Taylor once noted, these conditionals all seem to be equally suitable candidates for the analysis, and all of them seem to be defective. That they were *equally* suitable suggests that no one of them is *the* correct analysis, but the truth of such conditionals may yet be relevant to the truth of the 'could have' statements. Each of the conditionals might entail that the person could have done otherwise. Suppose a conditional statement *H* entails a 'could have' statement *C*. If, as is assumed, the conditional statement *H* is logically compatible with the truth of determinism, then it follows that the 'could have' statement *C* entailed by *H* is also logically compatible with determinism. For, it is a general theorem about consistency that if *P* entails *Q* and *P* is compatible with *S*, then *Q* is compatible with *S*.

Thus, if any single conditional statement of the sort in question entails a 'could have' statement, that would suffice to establish the compatibility of freedom and determinism on the assumption that the conditional is compatible with determinism. However, the arguments against the conditional analysis suggest the analysis fails because the 'could have' statement might be false when the conditional statement is true. This means that the conditional statement fails to entail the 'could have' statement. What is worth noting, however, is that a *conjunction* of a set of conditionals might yet entail the 'could have' statement. If the conjunction of conditionals did entail the 'could have' statement and if that conjunction was also compatible with

Peter van Inwagen (ed.), Time and Cause, 187–201.
Copyright © 1980 *by D. Reidel Publishing Company.*

determinism, then so would the 'could have' statement be compatible. Thus, to refute compatibilism, it will not suffice to point out that individual conditional statements do not entail the 'could have' statement. It is also necessary to prove that no conjunction of conditionals that is compatible with determinism entails the 'could have' statement.

The foregoing observation is not a mere logicial possibility, moreover. I shall attempt to show that a certain conjunction of conditionals entails the 'could have' statement. Let us note, first of all, the reason for thinking that the conditional analysis is plausible, for example, for thinking that '*S* could have done *A*' is analyzable as '*S* would have done *A* if *S* had chosen to do *A*'. The primary reason for finding such analyses plausible is that they capture one essential feature of freedom: that there must not be anything in the environment that prohibits the agent from doing *A*. Thus, if the agent is physically constrained, locked in a cell, chained, or held captive by others, then there will be certain actions he cannot perform. Such constraining circumstances insure that the corresponding conditional is also false. If I am held on the ground by several large and muscular men, it is false that I could have assisted a friend some distance away, and it is also false that I would have assisted my friend if I had chosen to do so. Indeed, most cases in which it is false that a person could have done something, are also cases in which it is false that he would have if he had chosen to do so. When choosing to do something would not have resulted in doing it, usually some circumstance prevents the person from performing the action. The circumstance need not be external to the body of the agent. Someone with a broken leg could not have run to the store, but neither is it true that the person would have run to the store if he had chosen to do so.

The problem with such analyses, and the reason they do not entail the 'could have' statements in question, is that some conditions would prevent a person from choosing to do something but would not prevent him from performing the action if he *did* choose to do so. Certain forms of compulsion provide the best examples. A person who has a severe obsessive phobia concerning blood may be unable to choose a red candy ball from a dish, may choose a green one instead, though, circumstances being what they are, the person would have obtained a red candy ball had he chosen to do so. Nothing would have kept the person from acting had he chosen to do so; it is, rather, that being the way he is, such a choice is out of the question. He cannot control himself well enough to choose in that way.

The reason the analysis fails is that the conditional may be true when there is some condition which renders the agent unable to fulfill the antecedent of the conditional and, therefore, renders the agent unable to perform the action. The conditional, though it has a false antecedent, is true even though the corresponding 'could have' statement is false. Thus, a compatibilist who wishes to appeal to conditional statements to prove his thesis must *supplement* the conditional statement considered above with some others to obtain a compound statement that entails the 'could have' statement. But what sort of additional statements should be added?

II. HIGHER-ORDER PREFERENCES

Recent work concerning preferences, especially higher-order preferences, provides the basis for an answer. Let us consider the case of someone who is compulsive. Frankfurt,[1] and subsequently Jeffrey,[2] suggest a characterization of compulsion in terms of second-order perferences. A person who is a compulsive smoker prefers to smoke when offered a cigarette. The person thus has a first-level preference for smoking over not smoking. Let us suppose that the person would rather not have such preferences because he believes that smoking is bad for his health. We put this by saying that the person has a second-order preference not to have the first-order preference to smoke. In other words, he prefers smoking to not smoking, but he also prefers not having that preference to having it. Thus, we have a distinction between, $P(S, S')$, preferring smoking to not smoking, and the following preference $P[P(S', S), P(S, S')]$ which is the second-order preference to have a first-order preference for not smoking over smoking. To have a simpler method of representation, let $P_1(S)$ represent a first-order preference for smoking over not smoking, $P_1(S')$ a first-order preference for not smoking over smoking, and let $P_2(S)$ represent the second-order preference for $P_1(S)$ over $P_1(S')$, $P_2(S')$ for $P_1(S')$ over $P_1(S)$, $P_3(S)$ for $P_2(S)$ over $P_2(S')$, and so forth so that $P_n(S)$ is the nth order preference for $P_{n\text{-}1}(S)$ over $P_{n\text{-}1}(S')$.

The foregoing symbolism naturally raises the question of whether higher-order preferences can be given any psychological meaning. Let us consider our smoker again. He prefers not to be the sort of person who prefers to smoke. It bothers him that he has that first-order preference. So he has a second-order preference not to have the first-order preference for smoking. Let us imagine that he has that second-order preference because he is very much preoccupied with his health. Let us further imagine, however, that he would prefer not to be so concerned about such matters, that he would prefer to be a person with more of a devil-may-care attitude toward matters of what is healthy and what is not. And, were he such a person, which he would prefer to be, he would smoke because he likes to smoke. So, our person has a second-order preference for not smoking because he is concerned about his health. But he has a third order preference not to have this second-order preference because he prefers to be a devil-may-care smoker. He has a third-order preference not to have the second-order preference not to have the first-order preference to smoke.

Is it possible to represent the person as having a fourth-order preference? Suppose that he is concerned about this preference to be a devil-may-care type of fellow; perhaps he thinks that this is something neurotic in him. Consequently, he prefers not to have the third-order preference for smoking because of his desire to be a devil-may-care type. Therefore, his fourth-order preference is not to have the third-order preference. And, it seems to me, that we might imagine that he wishes he did not have this qualm about being neurotic, and, therefore, has a fifth-order preference not to have the fourth-order preference. And so on. Represented in our symbolism, the preferences of the person would be: $P_1(S), P_2(S'), P_3(S), P_4(S'), P_5(S)$, and so forth. The person we have described is one that is full of psychological conflict, and such conflict is the stuff of psychological dysfunction. It is problematic whether when the person confronts the choice to smoke and chooses to do so, he was able to desist; his inner conflict may deprive him of control over his choice. The same may be said of him when he desists from smoking. When he desists, the preferences that lead him in that direction may undermine his control over how he chooses. With the inner turmoil we have imagined, he has relatively little rational control.

III. PREFERENCES AND CONDITIONALS

The seasoned campaigner for compatiblism will discern the solution to a traditional compatibilist problem in the account of higher-order preferences. We noted earlier that the attempt to analyse '*S* could have done *A*' as '*S* would have done *A* if *S* had chosen to do *A*' fails because of possible internal obstacles to choosing to do *A*, for example, when the person compulsively avoids that choice. Another way of putting the matter is simply to ask whether the person could have chosen to do *A*. If not, then it may be that the person could not have done *A*, because he could not have chosen to do *A*, even though he would have done *A* if he had chosen to do *A*. If, on the traditional analysis, we substitute 'choose to do *A*' for '*A*' to determine whether the person could done *A*, we get the result that the person could have chosen to do *A* if and only if he would have chosen to do *A* if he had chosen to choose to do *A*. Now some detractors have had their doubts about the meaningfulness of speaking of choosing to choose to do *A*. Moreover, even if that is conceded to have a clear meaning, the question may arise as to whether the person could have chosen to choose *A*. The answer on the traditional analysis is that the person could have chosen to choose to do *A* if and only if the person would have chosen to choose to do *A* if the person had chosen to have chosen to choose to do *A*. Doubts about the meaninfulness of talking about having chosen to have chosen to choose to do something are naturally intense.

Such doubts about nested choice may be avoided by appeal to nested preferences, for the latter have already been given interpretation. Consider the result of analyzing 'could have' statements in terms of preferences rather than choices. Suppose '*S* could have done *A*' were analyzed as '*S* would have done *A* if *S* had preferred to do *A*'. Now consider the question of whether the person could have preferred to do *A*. To say he could have would then mean that the person would have preferred to do *A* if he had preferred to prefer to do *A*. Since the conception of preferences among preferences does make sense, such an analysis is intelligible. But is it plausible?

I choose to drink a gimlet. Could I have desisted? Suppose I would have desisted if I had preferred to. Could I have preferred to desist? Suppose, in fact, that I did prefer to have the preference to desist, but lacking control, I actually preferred to drink. So, I preferred not to have the preference I acted upon, but the preference remained nonetheless. In that case, I could not have desisted because of psychological dysfunction. Thus, the analysis is too broad.

Now let us consider a second case. Again I choose to drink a gimlet. I prefer to drink in the circumstances. Moreover, there is no conflict, I am content with that preference. I prefer to have the preference I do have. If, moreover, I had preferred not to have that preference, then I would not have had it. If I had preferred to prefer to desist, I would have preferred to desist. I avoid conflict by integrating lower-level preferences with higher-order preferences. Imagine this integration occurs at every higher level.

The sort of integration of preferences just described is not a necessary condition for saying that a person could have desisted. The analysis is too restrictive. For we may wish to say of a person whose preferences contain some conflict at some level that he could have performed the action nonetheless. The thesis I wish to advance, however, is that such perfect integration of preferences provides a sufficient condition for saying that a person could have done otherwise.

Such total integration of preferences may be represented in terms of the symbolism introduced earlier. We may say that *S* is totally integrated with respect to doing A if and only if for any level i, if S had the preference $P_{i+1}(A)$, then S would have the preference $P_i(A)$. A person who is totally integrated with respect to desisting, as in the previous example, is such that if the person had preferred to have the preference to desist, then he would have had the latter preference. And if he had preferred to prefer to have the preference to desist, he would have preferred to have the preference to desist. And so forth up to higher orders of preference.

IV. A SUFFICIENT CONDITION FOR FREEDOM

Let us now state a condition that is sufficient for saying that a person could have done A when, in fact, he did not do so. Two conditions should be met: that the person would have done what he preferred to do, and, secondly that his preference would have been totally integrated. Then he could have done otherwise. In order to avoid the possibility that the person does not have higher-order preferences, we stipulate that the subject has higher-order preferences. We shall put this by saying that if $P_i(A)$, then either $P_{i+1}(A)$ or $P_{i+1}(A')$ or $I_{i+1}(A)$, where the latter says that the subject is indifferent about his ith-level preference concerning A. It appears necessary to introduce the notion of indifference as a primitive for our purposes rather than simply defining $I_{i+1}(A)$ as meaning that it is not the case that $P_{i+1}(A)$ or $P_{i+1}(A')$, because the latter is compatible with the subject not having an $(i+1)$-level preference in the way in which a river does not have a preference. In short, we wish to distinguish between being indifferent between alternatives when one might have had a preference and not having any preferences when having a preference was not an option.

With these distinctions before us, we can formulate our thesis more precisely as follows:

(i) S has a hierarchy of preferences, that is, if $P_i(A)$ then either $P_{i+1}(A)$ or $P_{i+1}(A')$ or $I_{i+1}(A)$.

(ii) S is totally integrated with respect to A, that is, if $P_{i+1}(A)$, then $P_i(A)$.[3]

(iii) S is not blocked from doing what he prefers with respect to A, that is, if $P_1(A)$, then S does A.

(iv) S does not do A.

Now conditions (i), (ii) and (iii) jointly entail that S could have done A, that is, S was free to do A.

There are a number of questions posed by these conditions that must be dealt with. First, we must defend the thesis that people sometimes have a hierarchy of preferences. Second, we must argue that a person might actually fulfill these conditions. Finally, to sustain compatibilism, we must show that the satisfaction of these conditions is compatible with the truth of determinism. We have laid the basis for this defense in our earlier discussion, but further argumentation is appropriate to bolster our claim against objections.

V. OBJECTIONS AND REPLIES

I am inclined to give the following defense of my thesis of the hierarchy of preferences. I prefer to write this paper. I prefer to have the latter preference. I also prefer to have the preference just mentioned, and so forth. I have no conflict at any level about writing this paper. I want to write it, I want to want to write it, I want to want to want to write it, and so on. I can go on as long as anyone might care to inquire. Moreover, if, contrary to what is the case, I preferred not to write this paper, I would have preferred not to have that preference. I am that way about writing and not writing papers, at least since I passed a certain stage of my career. I write them when I prefer, I do not write them when I prefer not to, and I no longer have any conflicts in such matters. If I prefer to prefer to write, then I prefer to write. If I prefer to prefer not to write, then I prefer not to write. When I make up my mind about whether to write or not, that settles the matter up and down the hierarchy of my preferences. If I am right about myself in this matter, then I need not advance any thesis about anyone else to sustain my thesis. Perhaps no one else has a hierarchy of perferences, though I believe that at least some do, and perhaps no one else is fully integrated in his preferences with respect to any action the person prefers to perform, though I suspect many are, but that does not matter. If the four conditions are ever satisfied for any person with respect to any action, then if he does not perform the action, he could have performed it nonetheless. If there is one such person satisfying those conditions with respect to one action, and if the satisfaction of those conditions is compatible with determinism, then freedom is compatible with determinism.

There is one serious line of objection to this thesis. It is that people could not possibly have higher-order preferences. Suppose a person allows that people could, conceivably, have higher-order perferences, but objects that as a matter of fact they do not. In that case, I would reply that if people could have such higher preferences, then conditions (i) to (iv) could be satisfied. They just happen not to be. If they could be satisfied, and if the satisfaction of those four conditions is compatible with determinism, then freedom and determinism are compatible. It might turn out that, as a matter of fact, there are no free actions, perhaps because people lack higher-order preferences and are, therefore, not much different from lower animals who lack freedom. But that would not be due to any logical inconsistency between freedom and determinism.

So the serious objection is one that says that we could not possibly have higher-order preferences. I believe we do have such preferences, as I have said. But further argument for the possibility of such preferences might be useful.

Note that if there are second-order preferences, it is difficult to find any reason for doubting the possibility of third and higher-order preferences. As a matter of fact, most people being unreflective about higher-order preference, they might lack such preferences. But if second-order preferences are allowed, then, by reflecting, a person might obtain preferences of any higher order. For second-order preference is a preference to have a preference. If it is possible that a person prefers to have one preference, it is possible that he prefer to have another. Once the possibility of preferences among preferences is conceded, it becomes possible that a person prefer to have, or not to have, any preference the person has, or does not have. The preference one prefers to have may be a second-level preference as well as a first, or an $(i+1)$-level preference as well as an i-level preference.

The question, then, is whether it is possible that people should have second-level preferences. An argument to the contrary, the only one of any merit, would contend that to prefer *A* to *B* is to choose *A* instead of *B*. Preference is marked in terms of choice. This view is behaviorism, and behaviorism has some detractors, but it could be the correct view in this special context. There is a traditional conception of preference in social science according to which a person prefers *A* to *B* if and only if, given a choice between *A* and *B*, the person chooses *A*. It might then be argued that what appears to be a second-order preference is really only an unusual first-order preference.

Consider the case of the person who was said to prefer not to smoke when, in fact, he chooses to smoke. The alleged second-order preference must be interpreted in terms of some choice the person would make. Suppose, for example, that a drug is available which will make a person vomit if he smokes and the alleged second-level preference not to have the first-level preference to smoke is strong enough so the person prefers taking the drug to not taking it, even though he knows that, for a little while anyway, he will still smoke. The alleged second-level preference, on the behavioristic interpretation, amounts to no more than that the person has the rather odd preference to smoke and vomit rather than to smoke and not vomit, and that is simply a first-level preference. In such a way, we can, according to the behaviorist, always account for alleged higher-level preferences in terms of genuine first-level preferences.

This line of thought may seem absurd initially. But it is possible to replace intuition with argument. First, we may concede to the behaviorist that second-order preferences must have some empirical implications to be scientifically respectable. But we need not concede that preferences are to be equated with choices. In the smoking case, for example, the person surely does *not* prefer smoking and vomiting to smoking and not vomiting. The second-order preference to prefer not to smoke may be manifested empirically in many ways without being equated with any specific choice the person would make. For example, the person may tell us of his second-order preference and show signs of anxiety when he lights up. He is certainly troubled about his smoking. To suppose that the attribution of the second-order preference has empirical content, it is not necessary to assume that the second-order preference is equivalent to any specific choice. Preferences should not be equated with the choices they explain because they explain more than simple choices. If preferences are not reducible to choices, then second-order preferences are not reducible to first-order preferences.

There is, moreover, a more compelling line of argument for second-order preferences. For it is quite beyond doubt that there are second-order preferences of an interpersonal variety. Suppose you prefer to smoke on an airplane when I am seated next to you and abhor cigarette smoke. I certainly prefer that you not have the preference you do have, that is, I prefer that you would prefer not to smoke. Now, if you are sensitive to my feelings and attitude toward you, you might prefer that I not prefer that you not prefer to smoke. You would rather that I did not mind. Now, next, I might prefer, not only that you desist from smoking, but that you also approve of people with my attitude toward smoking, that is, I prefer that you prefer that people have preferences like mine toward smoking and smokers. So, I prefer that you prefer not to smoke. I am somewhat evangelical about nonsmoking. The story could be extended. The crux is that higher-order interpersonal preferences do exist, and, indeed, account for many conflicts, and perhaps some felicities, in human relations. If I have preferences about your preferences, surely I also have preferences about my preferences.

VI. COMPATIBILITY OF FREEDOM AND DETERMINISM

Let us suppose, then, that it sometimes does happen that conditions (i) to (iv) are satisfied, that a person does not perform *A* when he has a hierarchy of preferences, his preferences are integrated with respect to A, and he is not blocked from doing A if he prefers to. My thesis is that the joint satisfaction of these conditions entails that a person could have done A and is, at the same time, logically compatible with the truth of determinism. To prove the theses, we need to establish two subtheses: first, that the joint satisfaction of conditions (i) to (iv) entails that the person S could have done A, that S was free to do A; and second, that the satisfaction of those conditions is logically consistent with the thesis that everything is determined.

That the joint satisfaction of the conditions entails that a person could have done A is sustained by consideration of the *control* a person has with respect to the action when the conditions are satisfied. A person may lack control over external circumstances in a way that prevents the person from performing the action. But suppose that a person has the required skills, there are no physical obstacles in his way, and he knows how to use those skills to do A. It might still be true that he could not have A, for he may lack control over himself rather than over the circumstances, and, for lack of such control, cannot perform the action. When a person has the control he requires to do A, then, even though he does not do A, he could have. The condition stating that the person would perform A if he preferred to do so insures that he does not lack the needed control over the external circumstanes. The condition that says if he preferred to have a preference, then he would have that preference, insures that he has control over himself. That is all the control he requires.

The conditional analysis, which says that if a person satisfies some condition C, then the person would do A, confronts the query – But could the person have satisfied C? Moreover, if that query were answered by saying the person would have satisfied C if some other condition were met, then the query will arise as to whether the person could have satisfied the antecedent of that conditional. And this line of query, the objector protests, will arise from conditional to conditional showing no condition has been provided, proving the person could have performed the action. Does our proposal meet this objection?

Condition (ii) of total integration answers all these questions at once. For any preference of level i, the question of whether the person could have preferred to do A at that level is answered by replying that the person would have preferred to do A at that level if he had preferred to have that preference to do A. Because of the generality of the condition, every query at every level is answered affirmatively. Thus the interrogator must admit it would be pointless to raise such questions because all such queries are answered at the outset by condition (ii). Thus, when the four conditions are satisfied the person has control both over his preferences and over doing what he prefers with respect to action *A*. So he could have done *A*. Granting this, let us turn to the final issue of whether the joint satisfaction of these conditions is compatible with determinism.

VII. DETERMINISM

The problem of how to define determinism is far from trivial. One method is in terms of possible worlds. By this method, assuming a notion of a logically possible world, we may say that determinism is true in the actual world if every possible world having the same laws of nature as the actual world and sharing a temporal slice with the actual world, that is, being identical to the actual world at some time, is identical to the actual world at every other time. This condition assures that, given the state of the world at any time, the state of the world is determined at every time by the laws. Thus other possible worlds having the same laws of nature as the actual world cannot differ from the actual world at any time if there is a single time at which they do not differ.

However, this condition, though sufficient, is not, I think, a necessary condition for a conception of determinism that says that for every event there is an antecedent sufficient condition, that condition itself having an antecedent sufficient condition, and so on into the indefinite past. This form of determinism must be true if the former is, but not vice versa. For it is possible that every event has antecedent sufficient conditions but some do not have subsequent sufficient conditions. The former definition requires that there be subsequent sufficient conditions as well in that it guarantees that what happens at one time is sufficient to determine both what happened previously and what will happen subsequently.

The second notion of determinism, I call it *antecedent* determinism, may be defined as follows. Antecedent determinism is true in the actual world if every possible world having the same laws of nature as the actual world and sharing a temporal slice with the actual world, that is, being identical to the actual world at some time, is identical to the actual world at every *subsequent* time. This insures that every event has antecedent sufficient conditions that recede as far into the past as one cares to trace them. We say that worlds that are identical at a time share a temporal slice at that time. Suppose that *E* occurs in the actual world, *E* is part of the temporal slice *T* at *t*. Now consider any antecedent time t' and the temporal slice T' at t'. Every world that shares

T' at t' with the actual world must, if antecedent determinism is true, share T at t with the actual world. So T' at t' in the actual world is a sufficient condition for T at t in the actual world, and therefore, for E at t in the actual world.

With these conceptions of determinism before us, we must confront the issue of compatibility. We have in (i) to (iv) a set of conditions that are, I believe, rather clearly consistent as a set, and the question is whether they are consistent with determinism as well. The logical analysis of the situation is as follows. We have a certain condition that fails to hold in the actual world. The person fails to do A. Then what is asserted in conditions (ii) and (iii) is a set of conditionals of the form: if A_1 had been the case, then A would have been the case; if A_2 had been the case, then A_1 would have been the case, and so forth, while (iv) says A was not the case. Condition (i) is a simple assumption about S that creates no additional problem for establishing compatibility. The problem of specifying the truth conditions for conditionals is by no means a settled issue in the philosophical literature. However, enough has been said to construct the sketch of an argument for the consistency of such a set of conditionals with determinism.

Take a single contrary-to-fact conditional: if S had preferred to do A, then S would have done A. This sort of conditional is rather plainly compatible with the accounts of determinism. The truth conditions for the contrary-to-fact conditional specify a set of possible worlds in which the antecedent is true and which are minimally different, maximally similar, or bear some proximity relation to the actual world. If the consequent is also true in those worlds, then the contrary-to-fact conditional is true, otherwise it is not. Now a world that differs from the actual world so that the antecedent of a contrary-to-fact conditional is true at some time in the possible world but not in the actual world does not share a slice with the actual world at that time. That every world having the same laws and sharing a temporal slice with the actual world should be identical to the actual world at every time is surely compatible with there being a possible world which differs from the actual world in such a way that the antecedent and consequent of the contrary-to-fact conditional are both true. Such a possible world need only differ from the actual world at all times or have different laws from the actual world.

Now let us consider the consistency of a set of contrary-to-fact conditionals with determinism. Assume that the set of conditionals is true so that there are possible worlds in which the antecedent and consequents are true and which bear the appropriate proximity relation to the actual world. For simplicity let us suppose that there is a *single* world in the set of possible worlds providing the verification of each conditional, for example, the closest possible world to the actual world in which the antecedent of the conditional is true. Hence we have the following:

W_1 for 'If A_1, then A';
W_2 for 'If A_2, then A_1';
W_3 for 'If A_3, then A_2',

and so forth, where the conditionals are the contrary-to-fact conditionals implied by (ii) and (iii), and the worlds W_i are the worlds that verify the conditionals. Now all that conditions (i) and (iv) might imply that is germane to the issue is that no W_i is the actual world. But the existence of a set of worlds which either differ from the actual world at all times or have different laws from the actual world is perfectly compatible with there being only one world that has the same laws as the actual world and shares a slice with it, the actual world itself.

Before concluding the case for compatibilism, it might be useful to pinpoint where the incompatibilist argument collapses given the considerations adduced above. The incompatibilist is wont to argue as follows. Suppose that determinism is true and that *S* does not do *A* at *t*. Then at any time, t', before the birth of *S*, there was some condition, perhaps the state of the universe at t', such that it is a consequence of that state and the laws of nature that *S* does not do *A* at *t*. Therefore, they conclude, if *S* could have done *A* at *t*, then *S* could have brought it about that either the laws of nature are different or the state of universe is different at t'. But obviously *S* could not have brought about either of these conditions. So, *S* could not have done *A* at *t*.[4]

The defect in the previous argument rests on the assumption that if *S* could have done *A* at *t*, then *S* could have brought about the conditions cited. If *S* had done *A* at *t*, then, of course, either the laws of nature would have been different or the state of the universe would have been different. But that is not to say that the person could have brought about these conditions. On the contrary, it is false that if *S* had preferred that either the laws of nature not be as they are or that the state of universe at *t* not be as it was, then one of these conditions would have been satisfied. The crux of the matter is that if the preferences of a person had been different, then, if determinism is true, either the laws would have had to be different or the antecedent state of the universe would have had to be different. Either of these might have had to be different, but it does not follow from the fact that one or the other would have had to be different that the person could have brought about those differences.

VIII. SUMMARY

We have argued that freedom and determinism are compatible by presenting a sufficient condition for affirming that a person could have done otherwise, articulated in terms of conditionals concerning preferences of the agent. If a person would have acted in a certain way if he had preferred to do so, would have preferred to do so if he had preferred to so prefer, and so forth, then the agent could have acted in the way in question, for there was no impediment to his so acting if he preferred to do so, and no obstacle to his so preferring, if he preferred so to prefer, and so forth. In short, when the conditionals in question are all true there can be no impediment, obstacle, or condition that prevents his acting. These conditionals are sufficient for saying that the person could have acted in the specified way. Moreover, the assumption and set of conditionals is compatible with determinism. If the set of conditionals and the other assumptions that the person has a hierarchy of preferences and does not perform the action jointly entail that the person could have performed the action, and, if, as we aver, the joint truth of those conditionals and assumptions is logically consistent with determinism, then, we conclude, so is the claim that the person could have performed the action he, in fact, did not perform. The principle employed here is that if one condition entails a second, and the first is logically consistent with a given thesis, then so is the second. Any thesis that is consistent with a set of premises is consistent with any conclusion deduced from those premises. Though we do not claim to have provided an analysis of 'could have' statements, we did provide a sufficient condition for the truth of such statements which is perfectly consistent with the thesis of determinism. Thus is the compatibility of freedom and determinism sustained.

University of Arizona and
Center for Advanced Study in the Behavioral Sciences

NOTES

* This paper was conceived at the Center for Advanced Study in Behavioral Sciences when I directed a Summer Institute on 'Freedom and Causality'. I am indebted to my codirector, Sydney Shoemaker, for the central idea of the paper, to the participants in the institute, especially Lita Furby and Hugh McCann, for critical comments, and to Kit Fine for several valuable discussions. I am also much indebted to the editor of this volume for his critical correspondence.

[1] Harry Frankfurt, 'Freedom of the Will and the Concept of a Person', *Journal of Philosophy* **68** (1971), 5–20.

[2] Richard Jeffrey, 'Preferences Among Preferences', *Journal of Philosophy* **71** (1974), 377–391.

[3] The editor of the volume has pointed out to me the need for a restriction on these conditionals. A person could be such that if he were to prefer to have some preference, then he would have it, but he might be such that if these preferences were to exist, they would result from external causation or manipulation. My intention is that the person should be such that if he preferred to have a preference and this was entirely an internal matter, that is, not externally caused, then he would, as a result, have the preference, and this would not be externally caused either. So, the conditional if $P_{i+1}(A)$, then $P_i(A)$ is to be interpreted as meaning of $P_{i+1}(A)$ without being externally caused, then $P_i(A)$ without being externally caused.

[4] Peter van Inwagen, 'The Incompatibility of Free Will and Determinism,' *Philosophical Studies* **27** (1975), 185–199.

TIMOTHY DUGGAN

HABIT

> Habit is . . . the enormous fly-wheel of society, its most precious conservative agent. It alone is what keeps us all within the bounds of ordinance, and saves the children of fortune from the envious uprising of the poor. It alone prevents the hardest and most repulsive walks of life from being deserted by those brought up to tread therein. It keeps the fisherman and the deck-hand at sea through the winter; it holds the miner in his darkness, and nails the countryman to his log-cabin and his lonely farm through all the months of snow. . . . On the whole, it is best he should not escape. It is well for the world that in most of us, by the age of thirty, the character has set like plaster, and will never soften again.
>
> William James[1]

In this paper, I shall put forward an account of what it is for a person to have the habit of doing *X*, where *X* stands for a kind of action, e.g., smoking, rather than a particular act, e.g., my smoking now. This account will serve as well to explicate cognate notions such as doing *X* habitually, being in the habit of doing *X*, and doing *X* from force of habit.

In *Action and Purpose*, Richard Taylor argues that the sense of 'can' relevant to discussions of human agency admits of no analysis. Normally, according to Taylor, the statement 'I can move my finger' as well as the statement 'I can refrain from moving my finger' are both true and known to be true. If anyone doubts the truth of either, there is a simple and straightforward manner of showing that such doubts are misplaced. "Whether or not I do move my finger is", Taylor says, " 'up to me' or, to use a more archaic expression, is 'something within my power'. And this is certainly a philosophically baffling expression which I feel sure no one can ever analyze; yet it *is* something that is well understood . . . we understand what it is for something to be in our power; and the fact that no one can *say* what it is is no disconfirmation of this".[2] Taylor, I believe, is mistaken. The account of habit offered here provides an analysis of and makes essential use of the concept of voluntary ability which is what Taylor has in mind by the 'can' of human agency.[3]

Peter van Inwagen (ed.), Time and Cause, 203–217.
Copyright © 1980 *by D. Reidel Publishing Company.*

Considering the important, indeed crucial, role that the concept of habit plays in much philosophical thinking, it is noteworthy that comparatively little has been done by way of analysis of this concept. Hume is a case in point of a philosopher for whom the concept of habit is central. Of custom or habit, Hume says, "wherever the repetition of any particular act or operation produces a propensity to renew the same act or operation, without being impelled by any reasoning or process of the understanding, we always say that this propensity is the effect of Custom".[4] From these and other remarks, we may conclude that for Hume:

> *S* has the habit of doing *X* ≡
> (1) *S* has done particular acts of kind *X* frequently;
> (2) The frequent doing of particular acts of kind *X* have produced (caused) the propensity to do particular acts of kind *X*;
> (3) When *S* does particular acts of kind *X*, he does so unthinkingly, nondeliberately, unreflectively, that is, "not impelled by any reasoning or process of the understanding".

This analysis of habit, as we shall see, is both too inclusive and at the same time too exclusive. John Austin is another philosopher who makes extensive use of the concept of habit. Austin defined positive law as a species of command, specifically, a command issued by a sovereign possessing both the ability, the opportunity, and the willingness to inflict evil in the event of noncompliance. The sovereign, in turn, is that person or group of persons whose commands the bulk of society *habitually* obey, and who in turn is not in the habit of obedience to anyone else.[5] Thus, for there to be a law, or a legal system, most of the people, most of the time, must habitually obey most of the sovereign's commands. Austin provides no general account of habitual behavior nor, specifically, an account of habitual obedience. Nevertheless, it would appear that for Austin if one *regularly* complies with the law, e.g., one requiring annual submission of an income tax return, this would be both necessary and sufficient for his *habitually* doing so. Similarly, if one *regularly* does not perform a kind of action prohibited by law, e.g., embezzlement, or rape, then one may be said to obey that law habitually even though one has neither the desire nor the opportunity nor, perhaps, the ability to violate the law. This stretches the notion of habitually doing something beyond all recognition.

H. L. A. Hart has pointed out many deficiencies in Austin's account of the nature of law.[6] In particular Hart argues that the habit of obedience fails to to account for continuity of authority to make law, and the *persistence* of laws over time. Hart also points out, using the example of paying one's taxes, that "this – even though regular – has not the unreflective, effortless, ingrained character of habit".[7] Again, Hart says, "To say of a person that he has a habit, e.g., of reading a newspaper at breakfast, entails that he has for some considerable time past done this, and that he is likely to repeat this behavior".[8] This latter observation on habitual action, though correct, will not suffice as an account of habit in general. It is, of course, true that many, perhaps all, of the things we do habitually we have done, more or less, frequently in the past and are likely to do again in the future. It is also true that many of the things we do habitually we do effortlessly and unreflectively, e.g., using a particular kind of knot when tying our shoes. This is at least part of what Hume had in mind in saying that in habitual behavior we are not, "impelled by any reasoning or process of the understanding". Indeed, this is in part what William James had in mind in saying, "Habit simplifies the movements required to achieve a result, makes them more accurate and diminishes fatigue. . . . Habit diminishes the conscious attention with which our acts are performed".[9] Nonétheless, many of the things we do habitually cannot be correctly characterized as *effortless* or *unreflective*. Two examples should suffice to make the point: (1) a person may be in the habit of lifting weights each morning for an hour – certainly not effortlessly, and (2) a person may have the habit of checking the locks on all doors and windows before retiring. Typically he does this with great care and attention – certainly not unreflectively. Thus Hume's and Hart's account of habit is too exclusive. They would rule out many things which are properly called habit, the exercise of which requires effort or attention or both. Hume's account is too inclusive in that it would include addictions, compulsions, phobias and the like. The argument that these should be distinguished from habits comes later on. Hart's account, if it is meant to provide necessary and sufficient conditions for acting habitually, is too inclusive as well. It would include, for example, breathing. The properties of being effortless, unreflective, having been done many times in the past, likely to be repeated in the future, are, for any normal adult anyway, true of breathing. Yet breathing is not something that we have the habit of doing.

I now want to say something about *not* doing things. There are many ways of not doing things, and ways of accounting for a person's not doing this or that. I am not speaking Mandarin Chinese now nor have I ever thus spoken, and this is because I lack the ability to do so. An acceptable answer to the question 'Why didn't he (doesn't he, won't he) do *X*?' is, very frequently, 'He can't do *X* because he lacks the ability to do *X*'. There are, however, many instances of a person not doing things which he has the ability to do. The following is a nonexhaustive list of kinds of such instances.

A person may be unable to do *X*, not because he lacks the ability, but because he lacks the opportunity to do *X*. An accomplished pianist may be unable to play the piano here and now simply because there is no piano. A person may fail to do *X* in the sense of neglecting to do what was expected of him, or he may try and fail. A person may omit doing *X* by leaving *X* out, as in consciously omitting salt from one's diet. I want to focus on one mode of not doing *X, viz., refraining* from doing *X*. An understanding of refraining is crucial to an understanding of habitual action.

Refraining from doing *X* entails not doing *X*, but the converse is not the case. I do not leap out of the window, or burst into song, but I do not refrain from doing these things. I have no inclination to do these things; my not doing them in no way requires that I restrain or curb myself from doing them. A confirmed smoker on a desert island without tobacco does not smoke though he may intensely desire to smoke, but he does not refrain from smoking. He does not refrain from smoking because in this instance he lacks the opportunity to smoke. I do not play the piano though I have ample opportunity to do so – there is a piano in my home – but I do not refrain from playing the piano. I simply lack the ability to play the piano. I suggest that

> *S* refrains (is refraining) at *T* from doing *A* (where *A* stands for a particular act of kind *X*)[10] iff:
> (1) *S* believes at *T* that he has the physical or mental ability to do *X*;
> (2) *S* believes at *T* that he has the opportunity to do *X*;
> (3) *S* is inclined at *T* to do *A*, i.e., in the absence of a consideration to the contrary, it is likely that *S* would do *A* at *T*;
> (4) *S* wills at *T* not to do *A*;
> (5) *S* does not do *A* at *T*.

Though this account of refraining may not capture all the nuances of the term as it is used in ordinary discourse, it is nonetheless an apt definition. Concerning condition (1), '*S* has the physical or mental ability to do *X*' is defined as follows: for a reasonable number of times, if *S* were to will to do a particular act of kind *X*, then simply given a reasonable opportunity, *S* would do that particular act of kind *X*; and willing to do a particular act of kind *X* is not a logically necessary condition for doing that particular act of kind *X*.[11]

Willing to do so is not a logically necessary condition of my putting a bullet in my head, or for my putting a bullet in the head of someone else. Indeed, I *could* do these things while asleep. On the other hand, willing to to so *is* a logically necessary condition of committing suicide or committing murder. Abilities to do these kinds of things are not simple physical or mental abilities (though they involve physical or mental abilities) but *voluntary abilities.*[12] It follows from the analysis of refraining that it need not be the case that one must actually possess the kind of ability of which the particular act one refrains from would be an instance. For example, a basketball player in order to throw the game may refrain from making foul shots provided he *believes* he has the ability to make such shots, even though, unbeknownst to him, he has suddenly lost the ability to do so. To avoid being tiresome someone may refrain from salting his conversations with quotations in ancient Greek even though, unbeknownst to him, he no longer has the ability to do so.

Concerning condition (2): it is, I suppose, unproblematic that often a person who has the ability to do X does not do a particular act of kind X simply because he realizes that he lacks the opportunity to do X. We would say of him that in those circumstances he was unable to exercise his ability. It is perhaps less evident that a person who lacks the ability to do X may nonetheless be said to have the opportunity to do X. For example, it may run counter to the linguistic intuitions of some to say of a person who possesses a piano but does not have the ability to play the piano that he, nonetheless, has the opportunity to play the piano. Indeed, I suppose that there is *a* sense of 'opportunity' in which we would say of him that he does not have the opportunity to play the piano (though he may have the opportunity to acquire the ability to play the piano). Nonetheless, the sense in which I use the term 'opportunity' (as opposed to merely believing that one has the opportunity) is intended to call attention to aspects of the circumstances in which a person finds himself rather than to properties of the person himself. One way we test for the presence or absence of an ability is to provide the subject with the opportunity to do X plus a reasonable incentive to do X. Sometimes it turns out that the subject clearly lacks the ability, though equally clearly he has the opportunity. It would be, I suppose, cruel to say of a person without legs that he has the opportunity to dance. It would be cruel but true; and cruel in this instance precisely because the person so obviously lacks the *ability* to dance.

It should be noted that condition (2) parallels condition (1) in that it allows for the possibility that a person may refrain from doing a particular act of kind *X* while believing falsely that he has the opportunity to do that act. For example, a person may refrain from eating candy believing that there is candy in the drawer while unbeknownst to him his wife has removed the candy.

Concerning condition (3), it is tempting to substitute the phrase '*S* wishes or desires to do *X*' for '*S* is inclined to do *X*'. But this, I think, would be a mistake since it would suggest that all refraining results in some kind of displeasure. But one can, after all, refrain from doing what one does not want to do, indeed, one can refrain from doing what one positively wants not to do. A person who becomes aware that he is boorishly dominating the conversation may, with a degree of satisfaction, refrain from doing so though he is inclined to do so. It should be added that although they very often are conscious, inclinations need not be conscious.

Concerning condition (4): I propose no complete analysis of willing though an analysis of the *ability* to will or volitional ability shall be proposed. Yet, as 'will' is used in this context, to will to do a particular act of kind *X* is to do that act intentionally, or to try to do it. Conversely, to will *not* to do *X* is intentionally not to do *X*, or to try not to do *X*. 'Willing not' is closely related to the archaic 'nilling', as in Thomas Hobbes, "Though a man have in every long deliberation a great many wills and nills".[13]

Condition (5) makes clear that refraining is an achievement word. One may try to refrain and fail not through any fault of one's own, but because of unintended and unforeseen consequences of what one does; Consider, for example, a man who resolves to cease upsetting his wife at social gatherings. He quite innocently mentions the color puce at a cocktail party which to his astonishment upsets his wife. He satisfies conditions (1) through (4), let us imagine, but not condition (5). It is false to say of him that he has refrained from upsetting his wife though he tried to do so. Conditions (1) through (4) of the analysis have to do with the agent's state of mind whereas condition (5) includes as well the agent's behavior and the circumstances the agent finds himself in.

We are now in a position to give an account of habit. I do not claim that this account captures all of the many uses of that term in ordinary discourse. Thus, the analysis should be regarded as a kind of theoretical definition. I do claim that the analysis provided is of a central use of the expression '*S* has the habit of doing *X*' – that use in which we speak of trying to break habits, falling into the habit of doing things, developing good habits, and so on. Some things commonly called habits, as we shall see, will not count as habits on my analysis. The analysis is as follows:

S has the habit of doing *X* (a kind of action), iff,

(1) *S* has done particular acts of kind *X* frequently, and the frequent doings of these particular acts partially explain *S*'s propensity to do *X*;
(2) *S* has the voluntary ability to do *X*;
(3) There is an act X^1 such that either X^1 is identical to *X* or to a segment of *X*; and if *S* were to refrain from doing a particular act of kind X^1, this would require a special act of attention on *S*'s part;
(4) *S* has sometimes done particular acts of kind X^1 not being consciously aware of what he was doing at time;
(5) *S* has sometimes done particular acts of kind X^1 being consciously aware of what he was doing at the time.

Concerning condition (1): we have some inclinations which are *natural* or *innate* as opposed to acquired or learned. Thus, the inclination to sleep, to eat, to defecate from time to time, and so on, would be natural or innate inclinations. My inclination to do these things now has *not* been caused by, nor is it partially explained by, my frequent doings in the past. I use the term 'propensity' (partly for nostalgic reasons) to refer to inclinations which have a particular etiology. Thus all propensities are inclinations, but the converse is not the case. To say that *S* has the propensity to do a particular act of kind *X* is to say that frequently, when the appropriate circumstances arise, *S* does or wishes to do that act, and this is partially explained by *S*'s having frequently done particular acts of kind *X* in the past.

Many philosophers, of course, have recornized the central importance of the process of habituation – that is to say, the genesis of habits. How we 'come by' some regularly repeated mode of behavior is crucial. Preanalytic assumptions about 'genesis' are at work when we identify this or that behavior as habitual, so much so that if it were proven that the genesis were otherwise (e.g., science-fiction cases of brain programming) we would question whether the conduct in question was indeed habitual. The ethical virtues are, Aristotle says, 'the child of habit'. A temperate man, for example, is one who has the settled habit of acting temperately; and this habit is produced in us by frequently doing temperate acts. This, for Aristotle, is true of all the ethical virtues and of the vices as well. Hume, as we have already noted, stressed the fact that it is the frequent doings of particular acts of kind *X* which produce in us the habit of doing *X*. John Dewey puts the point this way, "habit-forming wears grooves; behavior is confined to channels established by prior behavior".[14] There are, as noted above, many things that we do frequently, even regularly, but not habitually. We inhale and we exhale, we eat and we sleep frequently and more or less regularly but not habitually. A person about to be deprived of all food and oxygen might ironically complain that he was in the habit of breathing and of eating but this would be regarded as the sorry joke that it is. Our first condition which contains the feature that it is the *frequent doings of particular acts of kind X* which partially explain the propensity to do *X* rules out such cases as eating and breathing as habits and, in general, all cases in which *S*'s doing *X* is a causal consequence of a physiological deprivation or satiation. Breathing, eating, sleeping, and so on, are examples of things we do because of physiological deprivations. A somewhat vulgar example of satiation would be a person who regularly, but not habitually, belches after each meal. Those inclinations, the presence of which can be explained without reference to prior doings, fall outside our concept of habit.

Condition (2) is to the effect that if *S* has the habit of doing *X* then *S* has the *voluntary* ability to do *X*. *S* has the voluntary ability to do *X* (a kind of action), if, and only if, *S* has the physical or mental abilities to do *X* and *S* has the ability to will to do *X*. *S* has the ability to will or volitional ability to do *X*, if and only if:

(1) *S* has the ability to believe that there are many and varied coercive incentives for doing any particular act of kind *X*, and almost always if *S* believed that any of these coercive incentives were present, he would will to do that particular act of kind *X*.

(2) *S* has the ability to believe that there are many and varied noncoercive incentives for doing any particular act of kind *X*, and for each of several of these incentives if *S* believed that it was present, he would, at least sometimes, will to do that particular act of kind *X*.

(3) *S* has the ability to believe that there are many and varied coercive incentives for *not* doing any particular act of kind *X*, and almost always if *S* believed that any of these coercive incentives were present he would will *not* to do that particular act of kind *X*.

(4) *S* has the ability to believe that there are many and varied noncoercive incentives for not doing any particular act of kind *X*, and for each of several of these incentives if *S* believed that it was present he would, at least sometimes, will *not* to do that particular act of kind *X*.

S has the ability to believe that *P*, if and only if, if *S* were presented with overwhelming evidence that *P*, then *S* would believe that *P*. And if *S* were presented with overwhelming evidence that not-*P*, then *S* would believe that not-*P*.[15] A coercive incentive is one which it would be unreasonable to expect *any* rational person *not* to act on. A noncoercive incentive is an incentive which is not coercive.

The analysis of the ability to will is, alas, complex, but complex concepts require complex analyses. First, a person may will to do that which he lacks the ability to will to do. Clear examples are to be found in the behavior of persons suffering from compulsions and other disabilities of the will. For example, a compulsive handwasher wills to wash his hands though he lacks the volitional ability to do so. He lacks the ability to will to wash his hands because there are no noncoercive incentives given which he would refrain from doing so for any great length of time. A person suffering from extreme claustrophobia, when threatened with death, or some other coercive incentive, may will to hide in a closet, though there are no noncoercive incentives given which he would will to do this. In short, the fact that a person does will to do something does not show that he has the ability to will to do that thing. Admittedly, it sounds paradoxical to say that a person wills to do that which he lacks the ability to will to do, but this is no more paradoxical than saying that he does that which he lacks the ability to do. A case of the latter would be where a person in driving a golf ball just happens to hit a bird in flight. Second, abilities, including volitional abilities, may come and go sometimes without our realizing it. Third, intending or resolving are distinguished from willing in that, unlike the case of intending and resolving, there is *no* temporal gap between willing and intentionally doing or trying to do. The same is true for willing not to do, or nilling.

The ability to will is crucial for an understanding of habitual action. Admittedly, we speak of habit-forming drugs, a drug addict's habit, of his $100 habit, of his kicking the habit, and so on. But, on the present account of habit, an addict will not have the habit of taking drugs. This is because he lacks the ability to will to take drugs, though he does, of course, will to take drugs. He lacks the ability to will to take drugs precisely because there are no noncoercive incentives given which he would will not to take drugs and this constitutes a disability of the will. The account of habit here proposed is intended to distinguish between habitual actions and actions which are the result of addictions, obsessions, compulsions and phobias. And this is one of its virtues. To say of a heroin addict that he has the *habit* of injecting heroin in his veins, or of an extreme claustrophobiac that he has the habit of avoiding entering small enclosed places, or of a person obsessed with thoughts of death, that he has the *habit* of thinking about death, is at best misleading. These unfortunate persons suffer from a disability of the will that goes far beyond what we think of as habitual behavior. Of course, the demarcation between habitual action and compulsive, phobic, or addictive behavior will not be precise, and there will be cases which we simply will not know how to classify, but this does not argue against the reality or importance of the distinction. A consideration in support of this distinction has to do with excuses. Both in informal and in formal (e.g., courtroom) contexts, the fact that an action is the consequence of an obsession, compulsion, phobia or addiction is an excusing, or at least a mitigating condition. This is so because these disabilities of the will diminish to some extent one's freedom and thus one's responsibility. The perpetrator can claim with some plausibility that he could not help doing it. This move is not an acceptable one in the case of habitual action. That is, it is no excuse to say, 'I couldn't have helped it, I'm in the habit of doing that sort of thing'. A further consideration in favor of distinguishing between habitual actions and actions due to addictions, compulsions, and the like, is that these latter cases, where the person suffers from a disability of the will, are cases of maladies.

Condition (3), we may recall, had to do with the special act of attention required in order to refrain from doing that which we have the habit of doing. Sometimes we have the habit of doing things that are very simple, e.g., putting the left shoe on before the right; sometimes we have the habit of doing things that are very complex, e.g., such intellectual capacities as spelling correctly or speaking grammatically. Other habits fall somewhere between, e.g., the habit of taking a walk each afternoon or the habit of smoking cigarettes.[16] Acknowledging the formidable difficulties involved in individuating actions, I shall allow myself to speak of acts and of segments or proper subsets of acts. I want to claim that if a person has the habit of putting on his shoes in a certain order then the act, the refraining from which requires a special act of attention, is *just that* – putting his shoes on in that order. So too for other very simple acts. But let us consider an act of moderate complexity, *viz*., taking an afternoon walk, and let that be an act of kind X. There is a segment X^1 of that kind of act which requires a special act of attention if it is to be refrained from. Or consider smoking. Smoking typically incolves, among other things, reaching for the pack, taking a cigarette, striking a match, holding the lighted match to the cigarette, puffing, and so on. A person who has the habit of smoking has the habit of doing all these things, not just the habit of puffing a lighted cigarette. Refraining from smoking involves refraining from some particular act which is a segment of the complex act *smoking*. It could be an initial segment, e.g., taking a cigarette from the pack, or a subsequent segment, e.g., holding a lighted match to the cigarette. In any case, it is this refraining which requires a special act of attention. In discussing conditions (4) and (5), I shall speak of particular acts of kind X' bearing in mind that what satisfies the variable X' *may* be identical with what satisfies X as it occurs in the analysandum and in conditions (1) and (2).

It is very hard to say just what this act of attention is, though I am convinced that all or most of us are acquainted with the phenomenon. In part, it involves becoming aware of what one is about to do, though this is by no means a complete characterization. In this connection Thomas Reid says the following. After distinguishing between instinct which he says is "natural", or, as we might say, innate, and habits, which are acquired, Reid considers and rejects a 'common' definition of habit, *viz*., "a facility of doing a thing, acquired by having done it frequently". This, Reid says, may be an adequate definition of what are called habits of art, that is, acquired skills such as playing the piano or riding a unicycle, but it will not do as a general account of habit. It will not do because, as Reid puts it, "the habits which may, with propriety, he called principles of action, must give more than a facility, they must give an inclination or impulse to do the action". Reid adds the observation,

> I conceive it to be part of our constitution, that what we have been accustomed to do, we acquire, not only a facility, but a proneness to do on like occasions; so that it requires a particular will and effort to forbear it; but to do it, requires very often no will at all. We are carried by habit as by a stream in swimming, if we make no resistance.[17]

The particular will and effort, the resistance, that Reid speaks of is what I call a special act of attention. If S has been in the habit of doing X but now it no longer requires this act of attention for him not to do particular acts of kind X^1, S has, as we say, broken the habit.

Concerning a condition omitted from the analysis, it is very tempting to suppose that if S has the habit of doing X, then refraining from doing particular acts of kind X^1 will result in displeasure for S. The temptation is in part a consequence of thinking of habits as 'bad' habits. Examples that frequently come to mind when reflecting on habitual behavior are such things as smoking and drinking, and it is very likely that a person who habitually smokes or drinks will suffer displeasure in refraining from smoking or drinking. The fact is, however, that we can refrain from doing many things we have the habit of doing without suffering any displeasure whatsoever. William James supplies a striking example. Virtually everyone who has the habit of brushing his teeth, brushes his teeth in a particular order, e. g., top outside, bottom outside, bottom inside, top inside. Yet it is easy enough to change this particular order, indeed to reverse it. Doing so need result in no displeasure, though a special act of attention will be required. One must attend to what one is doing, perhaps recite to oneself: 'First, top inside; next, bottom inside', and so on. Similar things could be said about simpler habits such as putting on one's right shoe before one's left shoe. It might be objected that refraining from doing what one has the habit of doing *must* result in displeasure however minimal. The displeasure could take the form simply of a mild sense of tension of disequilibrium, but it must occur if we are dealing with a *bona fide* habit. But to insist that there must occur displeasure, however mild, in the kinds of cases we are considering is to make this claim uninterestingly analytic, or to invoke the problematic notion of unconscious displeasure to handle troublesome cases.[18]

Any adequate analysis of *habit* would have to take account of the distinction between the general case of acting habitually and the special case of acting from 'sheer force of habit' or 'pure or blind habit'. Many writers have stressed this later aspect of habitual action. For example, Gilbert Ryle in attempting to draw a sharp distinction between habits and other acquired dispositions, particularly intellectual capacities, says,

> When we describe someone as doing something by pure or blind habit, we mean that he does it automatically and without having to mind what he is doing. He does not exercise care, vigilance, or criticism .... It is of the essence of merely habitual practices that one performance is a replica of its predecessors.[19]

Again, he characterizes actions done "from force of habit" as being done "in complete absence of mind"[20] and adds that "[the subject] may, after the act is over, be quite unaware that he has done it ".[21] The expression 'not being consciously aware of what he was doing at the time' that occurs in condition (4) of the analysis is meant to express what is intended by 'from sheer force of habit', 'from pure of blind habit' as well as such terms as 'unthinkingly', 'unreflectively', 'automatically', and so on, which are often used in characterizing habitual action. This is, of course, a central element in habitual behavior. For it to be true that S has the habit do doing X, it must be the case that sometimes S does particular acts of kind X^1 without being aware of what he is doing at the time. It may seem implausible in the case of complex kinds of actions that a person could perform them while not being aware of what he was doing at the time, but recall that X^1 need not be identical to the complete act of kind X but merely to a segment of such an act. Our familiar smoker may, as it were, suddenly find himself about to put a lighted match to his cigarette or he may find himself actually smoking a lighted cigarette. And this is a familiar phenomenon. Still, to think of acting from sheer force of habit as the exclusive or even the primary aspect of habitual action is a mistake. After all, acting from *sheer force* of habit is meant to be contrasted with acting habitually. It would be a mistake to obliterate that distinction.

On the other hand, suppose that a person regularly and over a long period of time does acts of kind X. Moreover, suppose that that person *never* does X^1 while being consciously aware of what he was doing at the time. He may, of course, become aware that he has done X^1, say, by being told. Or he may become convinced that he will do X^1 in the future on the basis of an induction. Yet he ends his days never having consciously engaged in X^1. Admittedly, it is hard to think of plausible examples of this kind. But simple reflex actions, e. g., blinking ones eyes or sneezing, where the action is straightforwardly a response to a stimulus might be cases in point. Condition (5) of the analysis is intended to exclude this kind of case from the domain of habitual action. In support of this condition, I would cite the conviction that habits are formed at least in part by intentional acts. And this is true both of those things we make it a habit of doing (usually good things) and those things we fall into the habit of doing (usually bad things).

A wise man once said: 'Ask not what you can do for your analysis, ask what your analysis can do for you'. The above analysis bears on a fundamental issue in moral education and education in general, an issue which I can only touch on here. R. S. Peters formulates what he calls "the paradox of moral education" as follows:

> given that it is desirable to develop people who conduct themselves rationally, intelligently and with a fair degree of spontaneity, the brute facts of child development reveal that at the most formative years of a child's development, he is incapable of this form of life and impervious to the proper manner of passing it on.. . . [nevertheless] they can and must enter the palace of Reason through the courtyard of Habit and Tradition. . . . However, it is often thought that, because of the very nature of habits, dwelling in the courtyard of Habit incapacitates a man for life in the palace of Reason.[22]

Peters 'resolves' the paradox by arguing that there is no incompatability between the use of intelligence and the formation or exercise of habits. And that those who think there is have been misled by assimilating all habitual actions to acting from 'force of habit'. But as he puts it, "Habits need not be exercised out of force of Habit".[23] A. C. Kazepides argues in a similar vein that

> the notion of a single-sense concept of habit [sheer force of habit] ... is mistaken; that in ordinary discourse the word 'habit' is used to designate a wide variety of activities ranging from Ryle's most rigid and unthinking single-track dispositions to the most flexable and intelligent higher grade disposition.

Thus he concludes that the teaching and learning of habits is a legitimate task of education.[24] The account of habit provided here, incorporating elements of the cognitive and the voluntary, will, I think, contribute to the dialogue between those who assert and those who deny this claim about the aims of education.[25]

Dartmouth College

NOTES

[1] William James, *The Principles of Psychology*, Macmillan, London, 1891, p. 121.

[2] *Action and Purpose*, Prentice-Hall, Englewood Cliffs, N.J., 1966, p. 55.

[3] The concept of voluntary ability was first introduced in Timothy Duggan and Bernard Gert, 'Voluntary Abilities', *American Philosophical Quarterly* 4 (1967), 127–135; reprinted in Myles Brand (ed.), *The Nature of Human Action*, Scott, Foresman, Glenview, Ill., 1970.

[4] *An Enquiry Concerning Human Understanding* 5.1.36, in *Enquiries*, ed. by L. A. Selby-Bigge, rev. by P. H. Nidditch, Oxford University Press, London, 1975, p. 43. Hume uses the terms 'custom' and 'habit' interchangeably though by far the most frequently used term is 'custom.'

[5] *The Province of Jurisprudence Determined*, Murray, London 1832.

[6] Herbert L. A. Hart, *The Concept of Law*, Oxford University Press, London, 1961, Chaps. 2–4.

[7] *Ibid*., p. 51.

[8] *Ibid.*

[9] *Op. cit*., pp. 112, 114.

[10] For example, *A* might stand for biting one's nails *here and now; X* would stand for bitting one's nails.

[11] This account of physical or mental ability is developed in Duggan and Gert, *op. cit.*

[12] While the concept of voluntary ability and the ability to will or volitional ability is introduced later, it should be noted that on this analysis one may refrain from doing that which one lacks the *ability* to will to do. Indeed, one may refrain from doing that which one *knows* one lacks the ability to will to do. For example, a person suffering from severe claustrophobia will refrain from entering small places even though on this or that occasion he may be inclined to enter such places. His refraining is due to his disability of the will and thus it is involuntary. This is why condition (1) has to do with physical or mental abilities rather than with voluntary abilities which include the ability to will or volitional ability.

[13] *Liberty, Necessity and Chance*, in *The English Works of Thomas Hobbes of Malmesbury*, ed. by Sir William Molesworth, John Bohn, London, 1840, Vol. 5, p. 362.

[14] *Experience and Nature*, open Court, Chicago, 1926, p. 280.
[15] A version of this analysis of the ability to will appeared in Duggan and Gert, *op. cit.* The present (much revised) analysis appears also in Bernard Gert and Timothy Duggan, 'Free Will as the Ability to Will', *Noûs* **13** (1979), 197–217.
[16] Smoking may, of course, be an addiction rather than a habit as is the case with drinking. I have in mind here the analogue of the social drinker – call him the 'social smoker'. He, unlike the addicted smoker, has the voluntary ability to smoke.
[17] *Essays on the Active Powers of the Human Mind* 3.3, ed. by Branch A. Brody, M.I.T. Press, Cambridge, Mass., 1969, p. 115.
[18] The point here relates to a similar point made earlier in connection with refraining.
[19] Gilbert Ryle, *The Concept of Mind*, Hutchinson, London, 1949, 0. 42.
[20] *Ibid.*, p. 91.
[21] *Ibid.*, p. 110.
[22] R. S. Peters, 'Reason and Habit: The Paradox of Moral Education', W. R. Biblett (ed.), *Moral Education in a Changing Society*, London, 1963, pp. 107–110.
[23] *Ibid.*, p. 113.
[24] 'The Concept of Habit in Education', *Educational Theory* **20** (1970), 57.
[25] Thanks are due to Bernard Gert for many helpful suggestions.

RAYMOND MARTIN

EXPLANATORY CONTROVERSY IN HISTORICAL STUDIES

A Case Study: The Classic Maya Collapse

The problem of explaining the collapse in the ninth century A.D. of the Classic Lowland Maya civilization is among the most celebrated puzzles of archaeology. The present remarks are a study of certain aspects of the explanatory controversy which this problem has generated. My objective is to support a partial answer to the question *Q*: How do historians show that one explanation of some event is a better explanation of that event than are competing explanations?[1] What follows is divided into three parts. The first is devoted to a discussion of *Q*. The second is a characterization of the Maya collapse, of the principal hypotheses which have been proposed to explain it, and of the main arguments for and against four of these hypotheses. Elements of these arguments will be numbered to facilitate cross reference later in the paper. The third part of what follows is my answer to *Q*.

I

The question *Q* is important in the philosophy of history. The overriding objective of the historian, insofar as he argues on behalf of his explanations, is to show that his explanations are better than the best available competing explanations. If he appears to do this much, then so long as there is historical interest in the events he attempts to explain, his explanations will be taken seriously by his fellow historians, regardless of how incomplete his explanations or how slender the evidential base upon which they stand.[2] If he does not appear to do this much, then his explanations are not likely to be taken seriously. Hence, answering *Q* is tantamount to describing the logic of explanatory controversy in historical studies.

Question *Q* is also a neglected question. Philosophers of history, when discussing the evaluation of historical explanations, have tended to focus on the question, 'How *might* historians show that a *sufficient* explanation is *justified*?', while they have tended to neglect the questions, 'How *do* historians show that one explanation, *whether partial or sufficient*, is *better justified*, or more generally, *better*, than competing explanations?' But when a historian shows that his explanation is better than competing explanations he does not typically show that his explanation also satisfies the most widely

Peter van Inwagen (ed.), Time and Cause, 219–235.
Copyright © 1980 *by D. Reidel Publishing Company.*

discussed philosophical standards for a justified, sufficient explanation. As a consequence, philosophical accounts of the justification of historical explanation usually fail to make contact with many of the most prominent features of explanatory controversy in historical studies. And such accounts are often rejected by historians on the grounds that they are unrecognizable from the viewpoint of actual historical practice.

My attempt to answer *Q* requires preliminary qualification in three respects. Firstly, I shall confine myself to the consideration of explanations which can be expressed in the form, that *p* explains that *q*, where '*p*' and '*q*' are replaced by complete sentences in the indicative mood and that *p explains* that *q* just in case that *p at least partially explains* that *q*. I shall use the term 'explanans', to refer to the sentence which replaces '*p*' in explanations of this form and the term 'explanatory fact', to refer to what the explanans asserts to be the case. I shall use the term 'explanandum', to refer to the sentence which replaces '*q*' and the term 'explanandum event', to refer to what the explanandum asserts to be the case.

There is no single sentence which all archaeologists would agree should replace '*q*' in an explanation of 'the collapse'. Disagreement of this sort is typical of historical studies. But, it does not affect the *form* of the explanations under discussion. Nor will it undermine any argument that I wish to advance below. However, it will simplify my exposition to assume that at the present time there is agreement among archaeologists that some particular sentence expresses those aspects of the collapse for which an explanation is sought and that this sentence is both true and adequately justified. When I refer to *the collapse* I intend to refer to the fact which this sentence asserts to be the case.

Secondly, the notion of 'competing explanation' is unclear. Alternative explanations of the same explanandum event are not necessarily competing explanations of that event. The explanans of two alternative explanations may each be included in the explanantia of a third explanation which is more nearly sufficient than either of the two taken separately.[3] And alternative explanations may each be adequately justified sufficient explanations of the same explanandum event, in case the event is overdetermined or in case the explanans of each alternative explanation focuses on a different temporal slice of the causal sequence which culminates in the event.[4] The notion of a competing explanation seems best understood as an epistemic notion. Perhaps two explanations of an event are competing explanations of that event *if* it is more likely, on the available evidence, that one or the other of them explains that event than that both explain that event. For present purposes, I will not pursue the analysis of 'competing explanation' further. It is easy enough, and sufficient for my purposes, to identify pairs of explanations which archaeologists have regarded as competing explanations of the collapse.

Thirdly, the overall relative quality of competing explanations is affected by various of their characteristics. I shall confine myself to the consideration of just two such characteristics: the degree to which an explanation is justified or supported by the available evidence, and the degree to which an explanation is complete or sufficient. I believe that the majority of argumentation over the relative merits of competing explanations in historical studies is over these two characteristics of explanations. Certainly this is true of the explanatory controversy over the problem of explaining the Maya collapse.

II

The Classic Period lowland Maya civilization was the product of over a thousand years of cultural evolution in the tropical lowlands of the southern half of the Yucatán peninsula, largely in what is now the Peten district of Guatemala. By the end of the eighth century A.D. the lowland Maya had achieved their highest peak of intellectual, social, and artistic development. Their writing system was the most sophisticated in the New World. They had made important discoveries in mathematics and astronomy. Their economic, religious, and social systems were sophisticated. Their architecture, pottery, and painting were and are aesthetically impressive. At least twenty major centers and scores of minor centers flourished in and adjacent to the Peten. Then suddenly their civilization began its rapid decline.

The decline can be dated fairly precisely because the Maya were enthusiastic monument builders and invariably inscribed calendric information on their monuments. For example, in the more important centers in the lowlands, rows of carved stone slabs, called *stealae*, were placed periodically in the stucco floors of the plazas, usually facing important temples. In 790 A.D. nineteen of the major centers erected *stealae*. By 889 A.D., a year when tradition dictated that *stealae* be erected, only three centers erected *stealae*. Other monumental construction halted just as abruptly. The last known Classic Period calendrical inscription of any sort was carved in 909 A.D. Archaeologists believe that within a period of from fifty to one hundred years many of the centers and their immediate surroundings were completely abandoned and in most of the remainder there was only desultory occupancy. The area was never significantly populated again. According to one archaeologist, "the features that have intrigued scholars and produced the debates on the topic of the Southern Maya decline are the suddenness of the collapse and, more importantly, the fact that it was correlated with a massive population loss over an enormous area".[5]

Traditional explanations of the collapse tend to stress a single explanatory fact. Depending on the fact they stress, they may be classified as either catastrophic or ecological or social explanations. Catastrophic and ecological explanations stress the interaction between the Maya and conditions of nature. Social explanations stress interactions among Maya or between the Maya and foreign groups. More recent explanations tend to combine elements of several of the traditional explanations.

Catastrophic explanations explain the collapse by postulating natural disasters over which the Maya had little control: earthquakes, hurricanes, climatic changes, and epidemics have each been proposed as the explanation of the collapse. According to the *earthquake explanation,*[6] the fact that there were severe earthquakes in the southern lowlands in the late Classic explains the collapse. The evidence usually adduced for this explanation consists of (1.1) certain aspects of the present ruined condition of Classic centers, and (1.2) the fact that in modern times severe earthquakes are common in the highlands immediately south of Peten.

The earthquake explanation has been criticized on two major grounds. The first is that the explanans is not a sufficient explanation of the collapse. It is argued that (1.3) no comparable region is known to have been permanently abandoned as a consequence of earthquakes and that (1.4) earthquakes are not now and, so far as is known, were not in Classic times as severe in the lowlands as in the adjacent highlands where indigenous societies have survived serious earthquakes down to the present day. The second criticism is that there is insufficient evidence for the truth of the explanans. While most critics concede that relevant aspects of the present ruined condition of the centers *could* be a consequence of earthquakes, they insist that (1.5) they could *as likely* be a consequence of other causes as well. According to Adams, "it is very difficult to detect the difference between earthquake damage, structural failure and deterioration after abandonment".[7]

Ecological explanations, while varying greatly in details, attribute the collapse to environmental deterioration caused by overpopulation and overly intensive agriculture, the latter of which is said to have led to loss of soil fertility, excessive growth of grasses (savannaization) or soil erosion. According to the *subsistence failure explanation*, the collapse is explained by two facts: the Maya were almost exclusively dependent for their food supply on forest swidden maize agriculture[8]; and there was a significant growth in population among the Maya during the Late Classic. According to this explanation, as Maya population increased in the Late Classic, the swidden cycle had to be shortened, thus robbing the soil of valuable nutrients and lessening the yield of each planting. With shortened cycles the forest did not develop sufficiently to choke out weed growth. Weeds had to be removed by hand rather than burned, a nearly impossible task for the Maya given the difficulties of hand weeding and the large areas of land under cultivation. Eventually, shortened cycles led to the development of grassy savannas which replaced much of the forest. When this happened, the land was virtually lost as cropland since the Maya had no plow to turn the soil.

In order to support the claim that the Maya were almost exclusively dependent upon forest swidden maize agriculture for their food,[9] archaeologists have appealed (2.01) to the fact that modern lowland Maya have a quasi-religious allegiance to such a system, (2.02) to the compatibility of population distribution patterns in and around Maya centers with what one would expect under such a system, and (2.03) to the belief that such a system was the most productive system available to the Classic Maya. The claim that there was a population boom in the Late Classic is supported by complicated arguments which depend essentially on (2.04) the nature and character of surviving house mounds, low stone platforms upon which the Maya built their houses, and their associated ceramic and refuse deposits.

The claim that the two facts expressed by the explanans of the subsistence failure explanation led to a failure of the Maya subsistence system is supported by (2.05) contemporary laboratory studies of soil chemistry and agricultural yield under circumstances presumed to have obtained for the Classic Maya, (2.06) observation of the operation of analogous systems in other parts of the world, and, to a lesser extent, (2.07) indications of malnutrition in Late Classic skeletal remains.

Contemporary critics usually concede several aspects of the subsistence failure explanation. They do not question that there was a Late Classic population boom in the lowlands, that this put a serious strain on the Maya subsistence system, or that this strain played some part in the collapse. For example, Sabloff and Willey, in a paper arguing for the likelihood and importance of external invasion as a partial explanation of the collapse, concede that

> it seems highly probable that the balance between agricultural man and nature in the Southern Maya Lowlands was a precarious one, one that the strain of over-population – for which there is good evidence in Late Classic Period times – might have seriously disturbed and one that an historical event, such as invasion by an external enemy, might well have upset.[10]

The major contemporary criticisms of the subsistence-failure explanation question the explanatory importance assigned to subsistence strain. Criticisms which question the *sufficiency* of subsistence strain as an explanation of the collapse employ a familiar strategy. (2.08) Areas are mentioned which were subject to population problems as serious as and an ecology no more favorable than that in the Peten, but which nevertheless did not collapse. Other criticisms go beyond questioning the sufficiency of subsistence strain and attempt to minimize its importance relative to other explanatory facts. These criticisms usually question either the extent to which the Maya depended upon forest swidden maize agriculture or the extent to which their practice of such agriculture had deleterious effects.

Critics have questioned the dependence of the Maya on forest swidden agriculture in a variety of ways. They have argued that (2.09) the Maya cultivated root crops and *ramon* [bread nut] in addition to maize. For example, Sanders argues that "recent surveys by the Tikal Project have demonstrated an extraordinarily close relationship between the densities of ramon and housesites that indicates beyond any reasonable doubt that the Maya did plant groves of ramon trees near their houses".[11] They have argued that (2.10) the Maya may have farmed their swamplands in addition to the forest.[12] Finally, some critics have argued that (2.11) forest swidden agriculture was probably supplemented by the less efficient grass swidden system and that (2.12) the Maya may well have imported some foodstuffs.

Some critics question the extent to which the Maya practice of forest swidden agriculture had a deleterious effect on the maize crop. While it is clear that a shortened swidden cycle would result in less yield per labor hour, it is claimed that (2.13) with an abundant labor supply it may nevertheless have resulted in a greater net yield of maize. Many critics also question whether a shortened swidden cycle in the Peten led to the development of savannas. They note (2.14) the absence of centers from present savanna areas and the fact that (2.15) modern grassy roads and town plazas in the Peten, even ones that have been established for many years, quickly revert to forest if unattended.

An additional criticism sometimes mentioned is that (2.16) Copan and Quiriqua, to the southeast of the Peten, were two of the earliest centers to cease building even though they were situated in areas so fertile and well watered that exhaustion of their agricultural resources seems highly unlikely.

Social explanations attribute the collapse to some sort of social dysfunction. Peasant revolt, intercenter warfare, external invasion, trade disruption, and even an unfavorable sex ratio have each been proposed as explanations of the collapse. According to the *peasant revolt explanation*, as proposed by J. E. S. Thompson, the peasant class deposed, and probably destroyed, the elite class, thus putting an end to elite-class activities, such as monument building.[13] Thompson argued, contrary to the opinion of most archaeologists, that the peasants remained to populate the centers, that is, that the Maya suffered a decline of civilization without a corresponding decline of population. Thompson's main argument that there was a peasant revolt is based on (3.1) certain damage to and alteration of monuments, which he argued are most plausibly explained as deliberate acts of vandalism against the elite class. In addition, Thompson argued for an interpretation of Maya sociopolitical structure according to which (3.2) there was economic and ideological motivation for peasant revolt and (3.3) the elite class lacked those powers of coercion which would have protected it from revolts from within. Thompson appealed to two sorts of evidence in defense of his claims for a Post-Classic population in the Peten: (3.4) artifacts and burial remains in the centers which he argued were deposited after the collapse; (3.5) the fact that "there was a considerable population in the region in the sixteenth century", composed of individuals who he claimed are *most reasonably* regarded as "descendants of the original peasant population of the ninth century".

When Thompson first proposed his explanation, it was debatable whether the lowland centers were virtually abandoned during the Late Classic. Few archaeologists believe this is debatable today. According to Sanders, "virtually all recent archaeological studies in the southern lowlands agree that there was a catastrophic population loss".[14] According to Culbert, it is "extremely unlikely that anything yet to be discovered will change the conclusion that the Classic Maya of the southern lowlands suffered one of the world's great demographic disasters. . . . population loss may well have been in excess of a million people within a single century".[15] Thus, one major criticism of Thompson's explanation is that (3.6) it fails to explain why the centers were depopulated. Critics also argue that there is insufficient evidence for the truth of his explanans. For example, some critics have urged that (3.7) the evidence of vandalism and that the elite class was unable to protect itself against internal revolt favors certain competing explanations as much as it favors a peasant revolt explanation.

According to the *invasion explanation,*[16] the collapse is explained by the fact that there was a foreign, probably Mexican, invasion of the southern lowlands in the Late Classic. Versions differ with respect to the identity of the invaders and the timing and locus of the initial invasion. The invasion explanation is supported by two sorts of evidence. Firstly, (4.1) there is physical evidence of a foreign presence at several lowland centers. This evidence consists largely of ceramic remains which, it is argued, suggest the introduction of foreign pottery traditions, and certain aspects of monument remains. An example of the latter is that unusual-looking people with waist-length hair and bone nose ornaments, both of which were unknown among the lowland Maya, are pictured on several *stealae* at Seibal. Secondly, (4.2) there is substantial evidence of a Toltec invasion of Yucatán, north of the Peten, prior to or contemporaneous with the collapse of the southern lowland civilization; some invasion theorists argue that "since Yucatán was obviously invaded at about this time, it would be strange if the south had been completely untouched".[17]

The invasion theory has been criticized as an insufficient explanation of the collapse on the grounds that it does not explain the depopulation which occurred. Although it is possible that the invaders killed off the indigenous population, critics argue that this is unlikely. (4.3) The invaders in the Yucatan did not kill off the population there, and as one invasion theorist concedes, "even remarkably savage invasions, given pre-industrial instruments of homicide, have not characteristically depopulated entire regions ... "[18] It is also possible that the invaders upset the delicate subsistence balance, thus precipitating widespread starvation and/or migration. (4.4) Widespread starvation might account for a decline in population, but seems to some critics unlikely as an explanation of the almost total depletion which took place.[19] In any case, such an explanation tends to reduce invasion to trigger status and throws the burden of the explanation on noninvasion factors having to do with the vulnerability of Maya society. It is possible that migration contributed to the population loss, but there is little evidence that widespread migration occurred. Sanders has suggested a migration into the Guatemalan highlands to the south on the basis of evidence drawn from population estimates for the highlands, linguistic differentiation, and archaeological remains.[20] But some critics have rejected his suggestion on the grounds that (4.5) if Sanders's hypothesis were true, there would be more evidence in the ceramic remains of such a migration than there actually is. Cowgill has theorized that there was a series of Mexican invasions, which affected the entire lowlands, lasting possibly as long as several generations, and which substantially reduced the lowlands population, after which the invaders established a capital at Chichén-Itzá, to the north of the lowlands, and forced resettlement of the remaining lowlands population to localities within easy reach of Chichén-Itzá.[21] But Cowgill concedes that there is little evidence for his explanation and claims at most that his explanation is consistent with the evidence which exists. In sum, it seems that invasion without mass killing and/or substantial migration seems to most archaeologists an insufficient explanation of the collapse. And there is little positive evidence for an invasion accompanied by mass killing and/or substantial migration.

Another difficulty facing the invasion theory is that (4.6) at many centers, the existing evidence of foreign presence does not support the claim that there was a foreign presence *before* the collapse. Thus, it is sometimes argued that the foreign presence can be interpreted as easily as a consequence as it can as a cause of the collapse. And at some centers, such as Tikal, critics claim that (4.7) a stronger case can be made that the collapse preceded the foreign presence than that the foreign presence preceded the collapse.

III

Almost all arguments in support of an explanation of the collapse may be regarded as of one or another of three types: those which are intended to increase or diminish the likelihood (1) that a particular explanans is true, (2) that a particular explanans (at least partially) explains its explanandum, or (3) that a particular explanans is a sufficient explanation of its explanandum.[22] *Positive* arguments of each type are intended to increase the likelihood of the relevant aspect *of a favored explanation. Negative* arguments of each type are intended to diminish the likelihood of the relevant aspect *of some nonfavored, competing explanation*. For example, a positive argument of type (2) is intended to increase the likelihood that the explanans of a favored explanation explains its explanandum; a negative argument of type (3) is intended to diminish the likelihood that the explanans of a nonfavored, competing explanation is a sufficient explanation of its explanandum.[23]

Positive arguments of type (1) are almost always attempts to show that certain data are best explained by facts the existence of which increases the likelihood that the explanans of a favored explanation is true.[24] Usually, this strategy consists of attempts to show that certain data are best explained by the explanatory facts postulated by the favored explanation; for example, the attempt to show that certain aspects of the present ruined Classic centers are best explained by the fact that there were severe earthquakes in the Peten in the Late Classic, or the attempt to show that certain damage to monuments is best explained by the fact that there was a peasant revolt of such and such a kind in the Late Classic.[25] However, sometimes this strategy consists of attempts to show that certain data are best explained by explanatory facts which differ from, but increase the likelihood of, those explanatory facts postulated by a favored explanation; for example, the best explanation of the prevalence in modern times of severe earthquakes in the highlands immediately south of Peten presumably postulates certain facts concerning relatively permanent aspects of the geology of the area which increase the likelihood that there were severe earthquakes in the Peten in the Late Classic.[26]

Negative arguments of type (1) attempt to show that there is no evidence for and/or evidence against the explanans of some nonfavored competing explanation. The claim that there is no evidence for an explanans is simply the claim that there is no good positive argument of type (1) for that explanans, for example, the claim that there are no data which are better explained by the explanans of the earthquake explanation than by the explanans of some competing explanation.[27] The argument that there is evidence against an explanans takes one or the other of two closely related forms. The first purports to show that certain data do not exist that likely would exist if the explanans were true; for example, the claim that there was a severe epidemic in the Late Classic has been criticized on the grounds that there is no evidence of Late Classic mass burials nor any traces of epidemic diseases in the recovered skeletal material.[28] The second purports to show that certain data exist that likely would not exist if the explanans were true. For example, the claim that the Maya were almost exclusively dependent for their food upon forest swidden maize agriculture has been criticized on the grounds that there is a close relationship between the densities of *ramon* trees and house sites which is best explained by the supposition that the Maya planted *ramon* trees near their houses.[29]

Positive arguments of type (2) are arguments to the effect that facts of the same sort as at least some of the explanatory facts expressed by the explanans have a lawful relationship to facts of the same sort as at least some of the facts expressed by the explanandum. In the debate over the Maya collapse, such arguments were usually not given; it was simply assumed that the explanans, if true, explained the explanandum. However, when such arguments were given, it was apparently because either the relationship between facts of the same sorts as the explanatory facts and the explanandum event was not a familiar one or because the explanatory relevance of the explanans to the explanandum had been challenged. When such arguments were given, they consisted either of claimed positive correlations between other instances of facts of the relevant kinds or of appeals to theoretical considerations, or of both. An example of the former is Sanders's argument, urged in support of the subsistence failure explanation, that recent evidence "from Africa and

Indonesia very strongly supports the position that there is an overall succession from forest to grass when population densities reach certain levels in swidden agriculture".[30] Since Sanders does not regard agricultural strain as a sufficient explanation of the collapse, his argument must be regarded as an attempt to show that the growth of savannas and consequent agricultural strain are a partial explanation of the collapse. Shimkin has argued on the basis of both comparative and theoretical evidence, the latter simply an application of aspects of the science of disease, that archaeologists have underestimated the destructive effects on the Maya of epidemic and endemic diseases.[31] Again, this seems an argument in support of a partial explanation of the collapse.

In the debate over the collapse, I found arguments which were clearly intended to be positive arguments of type (2), but none which were clearly intended to be positive arguments of type (3). No one *explicitly* claimed, for example, to be showing that some explanans was a sufficient explanation of the collapse or of any other explanandum. On the other hand, negative arguments of type (3) were quite common, whereas negative arguments of type (2) were almost nonexistent.[32] This is just what one would expect of cautious archaeologists. It is easier to show that some explanans is a partial, rather than a sufficient, explanation of its explanandum. And it is easier to show that some competing explanans is not a sufficient, rather than not a partial, explanation of its explanandum. Since positive arguments of type (3) and negative arguments of type (2) played no important role in the debate over the collapse, I shall not discuss them below.

Negative arguments of type (3) were of one or another of three sorts. One sort argued from the premise that although facts of the same sort as the explanatory facts often occur, they have never been known to bring about events of the same sort as the explanandum event. For example, explanations which postulate earthquakes or hurricanes as the explanation of the collapse are criticized as insufficient on the grounds that although earthquakes and hurricanes are quite common, their known effects have always been local and sporadic, never the decimation of an entire civilization.[33] A second and closely related objection to the sufficiency of an explanation is that its explanans does not explain the difference between the situation in which the explanandum event occurred and some specific comparison. For example, Andrews questioned the sufficiency of the subsistence failure explanation on the grounds that although the Rio Bec area probably had an ecology which was virtually identical to that of the Northern Peten, only fifty miles to the south, "the dense populations of the Rio Bec continued successfully for a number of centuries after the Northern Peten centers were largely abandoned".[34] Finally, the sufficiency of an explanation is often questioned on the basis of what might be called theoretical considerations; for example, contemporary laboratory studies on soil depletion are sometimes cited in support of the claim that the forest swidden system of agriculture practiced by the Maya was not as hard on the soil as some have claimed. More informally, nutritional disease has been rejected as a sufficient explanation of the collapse on the grounds that "the effect would be a reduction of population to the point where a viable subsistence system would become reestablished . . ."[35]

I am now prepared to suggest a partial answer to *Q*. My answer is partial because I shall consider just the degree to which competing explanations are justified by the available evidence and the degree to which they are sufficient. Hence, *Q* reduces to three more specific questions: *Q*1: How do historians show that one explanation of some event is a better-justified explanation of that event than are competing explanations?; *Q*2: How do historians show that one explanation of some event is a more nearly sufficient explanation of that event than are competing explanations?; *Q*3: What is the relationship between showing that one explanation of some event is a better justified and/or more nearly sufficient explanation of that event than are competing explanations and showing that it is a better explanation of that event than are competing explanations? My answers to *Q*1 and *Q*2 will be descriptions of the primary argumentative strategies archaeologists have used to show that some explanation of the collapse was better justified and/or more nearly sufficient than competing explanations. My answer to *Q*3 will be relatively independent of the data generated by the present case study.

*Q*1: Archaeologists showed that one explanation of the collapse was a better-justified explanation of the collapse than were competing explanations primarily by showing, via a combination of positive arguments of sort (1) and negative arguments of sorts (1) and (3), that it is more likely that the explanans of a favored explanation is true than that the explanans of a competing explanation is true. The vast majority of positive arguments for the justification of an explanation consist of the presentation of data which are claimed to be best explained by the explanans of that explanation. The vast majority of negative arguments for the justification of an explanation question either the sufficiency of non-favored competing explanations or the truth of their explanans. The former is done primarily via the three strategies specified above in the discussion of negative arguments of type (3). The latter is done primarily by correcting data upon which such explanantia have been supported so that the data no longer support them as well as they did, and suggesting alternative more congenial ways in which such data as survive might be explained. A secondary, and rarely employed, way in which archaeologists of the collapse showed that one explanation of the collapse was a better-justified explanation of the collapse than were competing explanations, was by showing, via a combination of positive arguments of type (2) and negative arguments of type (3), that it is more likely that the explanans of a favored explanation explains the collapse than that the explanans of some nonfavored competing explanation explains the collapse.

*Q*2: Archaeologists showed that one explanation of the collapse is a more nearly sufficient explanation of the collapse than are competing explanations almost exclusively by showing via negative arguments of type (3) that nonvavored competing explanations are not sufficient and by adding to the explanans of the best justified of these explanations explanatory facts which constitute the best-justified relevant addenda available. For example, although most contemporary students of the collapse seem to subscribe to some version of the subsistence failure explanation, they also recognize that it does not account adequately for the depopulation which occurred. Thus, they add to the subsistence failure explanation an addendum to account for the depopulation. For example, Sanders argues that the explanans of his version of the subsistence failure explanation should be supplemented with information concerning a migration from the southern lowlands to the highlands south of Peten.[36] Sanders does not present impressive evidence for his addendum. But he feels the need for some addendum. And he argues that his addendum is better justified than the most plausible competing addenda. One can easily *imagine* alternative strategies a historian might employ to argue that his favored explanation of some event is more nearly sufficient than are competing explanations. But in the debate over the collapse such strategies are barely discernible. Thus, the kinds of arguments which are used in the debate over the collapse to support the claim that an explanation is more nearly sufficient than a nonfavored competing explanation are quite similar to the kinds of arguments which are used to support the claim that one explanation is better justified than nonfavored competing explanations. In both cases, they are mainly arguments supportive of the explanans, or of an addendum to the explanans, of a favored explanation and arguments critical of the justification or sufficiency of nonfavored competing explanations.

*Q*3: The notions of one explanation's being *better justified* than another and *more nearly sufficient* than another seem to me clearer notions than the notion of an explanation's being simply *better* than another. Still, the latter notion cannot simply be dismissed. Assume that E_1 and E_2 are competing explanations of the same explanandum event and that, except for the degree to which each is justified and the degree to which each approximates a sufficient explanation of their common explanandum event, neither is preferable to the other in any respect relevant to which is a better explanation of their explanandum. Given this assumption, the following two claims seem true: (1) if E_1 is better justified than E_2, and E_2 is not more nearly sufficient than E_1, then E_1 is a better explanation of their common explanandum than is E_2; (2) if E_1 is more nearly sufficient than E_2, and E_2 is not better justified than E_1, then E_1 is a better explanation of their common explanandum than is E_2. From (1) and (2) it follows that if E_1 is both better justified and more nearly sufficient than E_2, then E_1 is a better explanation of their common explanandum than is E_2.

The decision whether to advocate a particular explanation of the collapse seems to involve a weighing, on the part of archaeologists, of the advantages of superior justification against the advantages of greater approximation to sufficiency. Sometimes, if an explanation is known to be insufficient, an archaeologist will advocate a more nearly sufficient but less well justified explanation which includes the former. Sometimes an archaeologist will prefer to advocate a better justified, but less nearly sufficient explanation. Within rather broad limits such judgments seem to reflect temperamental characteristics of archaeologists rather than allegiance to some systematic method for making such decisions.

I have described how archaeologists have argued that some favored explanation of the collapse is better justified and/or more nearly sufficient than nonfavored competing explanations and I have used that description as a basis for suggesting a partial answer to Q. I believe, but have not shown, that the sorts of arguments and argumentative strategies that archaeologists of the collapse have employed are used quite generally in historical studies. If they are, then the discussion above suggests two important reasons why philosophical accounts of the justification of historical explanation so often look like a distorted caricature of actual historical practice. Firstly, such accounts typically underestimate the importance of justifying the explanans and overestimate the importance of justifying covering generalizations. The explanantia of those explanations which are the foci of most explanatory controversies in historical studies are almost always subjected to vigorous criticism; that such explanantia at least partially explain their explananda, given that both explanantia and explananda are true, is a less frequent topic of debate. Secondly, such accounts typically all but ignore the comparative evaluation of competing explanations. It is characteristic of explanatory controversy in historical studies that one argues for a favored explanation not only via positive arguments but also via arguments which reduce the plausibility of competing explanations. Whether or not the sorts of arguments and argumentative strategies that archaeologists of the collapse have employed are used quite generally in historical studies, if *Q* is ever to be answered adequately, it will be answered by means of the detailed examination of a number of explanatory controversies in historical studies and careful generalization on the basis of these case studies. The present paper constitutes one such case study.

University of Maryland,
College Park

NOTES

[1] I am using 'history', 'historian', and related expressions in a broad sense to include not only history proper but also archaeological enquiries of the sort under discussion in the present paper.

[2] Thus, it is a commonplace of historiography that historians will continue to debate the relative merits of competing explanations of some event even though none of the competing explanations is supported by substantial evidence.

[3] See, for example, W. T. Sanders, 'The Cultural Ecology of the Lowland Maya: A Reevaluation', in T. P. Culbert (ed.), *The Classic Maya Collapse*, University of New Mexico Press, Albuquerque, 1973, pp. 325–365, pp. 362–365.

[4] An example of the former may be found in Carl G. Hempel, *Aspects of Scientific Explanation*, Free Press, New York, 1965, p. 419; and of the latter in Raymond Martin, 'Marc-Wogau and Mackie on Singular Causal Statements', in *Philosophical Forum* 3 (1971), 145–151, p. 150.

[5] Sanders, *op. cit.*, p. 361.

[6] (1.1)–(1.2): S. G. Morley and G. W. Brainerd, *The Ancient Maya*, Stanford University Press, Stanford, Calif., 3rd ed., 1956, p. 69; E. W. MacKie, 'New Light on the End of Classic Maya Culture of Benque Viejo, British Honduras', *American Antiquity* 27 (1961), 216–224; R. E. W. Adams, 'The Collapse of Maya Civilization: A Review of Previous Theories; in Culbert, *op. cit.*, pp. 21–34, p. 27. (1.3)–(1.5): Morley and Brainerd, *op. cit.*, p. 69; G. L. Cowgill, 'The End of Classic Maya Culture: A Review of Recent Evidence', *Southwestern Journal of Anthropology* 20 (1964), 145–159, p. 152; Adams, *op. cit.*, p. 27.

[7] Adams, *op. cit.*, 27.

[8] In forest swidden maize agriculture, the forest is cleared with hand tools, the resulting debris burned, and maize planted for a period of time from one to three years. The land is then allowed to rest for several years before recultivation.

[9] (2.01)–(2.03): Adams, *op. cit.*, 24–26; T. P. Culbert, *The Lost Civilization*, Harper

and Row, New York, 1974, p. 46. (2.04): Elizabeth P. Benson, *The Maya World*, Thomas Y. Crowel, New York, 1967, p. 126; Sanders, *op. cit.*, pp. 328–332; Culbert, *op. cit.*, pp. 41–46. (2.05)–(2.07): Adams, *op. cit.*, pp. 26, 29, 32; Sanders, *op. cit.*, pp. 332–341. See also Cowgill, *op. cit.*, p. 153, F. B. Saul, 'Disease in the Maya Area: The Pre-Columbian Evidence', in Culbert (ed.), *op. cit.*, pp. 301–324; and G. R. Willey 'Certain Aspects of the Late Classic to Postclassic Periods in the Belize Valley', in Culbert (ed.), *op. cit.*, pp. 93–106, p. 100. (2.08): Cowgill, *op. cit.*, p. 152; E. W. Andrews IV, 'The Development of Maya Civilization After Abandonment of Southern Cities', in Culbert (ed.), *op. cit.*, pp. 243–265, pp. 259–260. (2.09)–(2.13): Adams, *op. cit.*, pp. 26; Culbert, 'The Maya Downfall at Tikal', in Culbert (ed.), *op. cit.*, pp. 63–92, pp. 71–72; Sanders, *op. cit.*, p. 340; G. R. Willey and D. B. Shimkin, 'The Maya Collapse: A Summary View', in Culbert (ed.), *op. cit.*, pp. 457–501, pp. 474–476, 482–483; Culbert, *Lost Civilization*, pp. 47–49. (2.14)–(2.16): Morley and Brainerd, *op. cit.*, p. 71; Cowgill, *op. cit.*, p. 152; J. E. S. Thompson, *The Rise and Fall of Maya Civilization*, University of Oklahoma Press, Norman, 1966, pp. 102–103; J. A. Sabloff and G. R. Willey, 'The Collapse of Maya Civilization in the Southern Lowlands: A Consideration of History and Process', *Southwestern Journal of Anthropology* **23** (1967), 311–336, p. 315.

[10] Sabloff and Willey, *op. cit.*, p. 316. See also Adams, *op. cit.*, p. 26, and C. J. Erasmus, 'Thoughts on Upward Collapse: An Essay on Explanation in Anthropology', *Southwestern Journal of Anthropology* **24** (1968), 170–194.

[11] *Op. cit.*, p. 340.

[12] For example, Culbert notes that prehistoric South Americans farmed their swamplands by "building ridged-field systems in which dirt is piled up to make ridges that are high enough to avoid destructive amounts of moisture [and that] . . . some [such systems] have been discovered along the Rio Candelaria at the western edge of the Maya lowland", in Culbert (ed.), *op. cit.*, p. 49.

[13] Thompson, *op. cit.*, pp. 100–109. See also Morley and Brainerd, *op. cit.*, pp. 57–73; Charles Gallenkamp, *Maya*: *The Riddle and Rediscovery of a Lost Civilization*, McKay, New York, 1959, pp. 158–163; Sabloff and Willey, *op. cit.*, pp. 317–318; Thompson, *Maya History and Religion*, University of Oklahoma Press, Norman, 1970, pp. 79–83, Adams, *op. cit.*, pp. 29–30; Sabloff, 'Major Themes in the Past Hypotheses of the Maya Collapse', in Culbert (ed.), *op. cit.*, pp. 35–40, pp. 37–38; Sanders, *op. cit.*, p. 346; and Willey and Shimkin, *op. cit.*, pp. 467–468, 485.

[14] *Op. cit.*, p. 361.

[15] *Lost Civilization*, p. 109.

[16] (4.1)–(4.2): Cowgill, *op. cit.*, p. 154; Benson, *op. cit.*, pp. 130–131; Sabloff and Willey, *op. cit.*, pp. 319–330; Adams, *op. cit.*, pp. 30–33; Adams, 'Maya Collapse: Transformation and Termination in the Ceramic Sequence at Altar de Sacrificios', in Culbert (ed.), *op. cit.*, pp. 133–163, pp. 149–158; Sabloff, 'Continuity and Disruption During Terminal Late Classic Times at Seibal', in Culbert (ed.), *op. cit.*, pp. 107–131, pp. 127–129; M. C. Webb, 'The Peten Maya Decline Viewed in the Perspective of State Formation', in Culbert (ed.), *op. cit.*, pp. 367–404, p. 402; Willey and Shimkin, *op. cit.*, pp. 464–465, 469–470. (4.3): Cowgill, *op. cit.*, p. 154. (4.4)–(4.5): Sanders, *op. cit.*, pp. 364–365; Willey and Shimkin, *op. cit.*, pp. 470–473. (4.6)–(4.7): Culbert, in Culbert (ed.), *op. cit.*, pp. 89–92; R. L. Rands, 'The Classic Maya Collapse: Usumacinta Zone and the Northwestern Periphery', in Culbert (ed.), *op. cit.*, pp. 165–205, p. 197. See also Webb, op. cit., p. 368.

[17] Cowgill, *op. cit.*, p. 154.
[18] *Ibid.*
[19] For example, Sanders argues that nutritional deprivation would result at most in "a reduction of population to the point where a viable subsistence system would become reestablished . . . ", (*op. cit.*, p. 364).
[20] *Ibid.*, pp. 364–365.
[21] Cowgill, *op. cit.*, pp. 155–156.
[22] There are, of course, also arguments which are intended to increase or diminish the likelihood that a particular explanandum is true. But since there is no difference in structure between these arguments and arguments of type (1), I consider subsequently just arguments of type (1).
[23] All arguments of type (1) support the conclusion of an argument of type (2). All arguments of type (2) support the conclusion of an argument of type (3) and vice versa. Hence, an argument in the archaeological literature may often be regarded as an instance of more than one of my argument types.
[24] (2.03) is one of the few clear exceptions which I discovered to this rule (Culbert, *Lost Civilization*, p. 46). See also the use made by D. B. Shimkin of what he calls "comparative evidence" in 'Models for the Downfall: Some Ecological and Culture-Historical Considerations', in Culbert (ed.), *op. cit.*, pp. 269–299.
[25] Other examples include (2.02), (2.04), and (4.1). See also (3.4) and (3.5).
[26] Other examples include (2.01), (3.2), (3.3), and (4.2).
[27] Other examples include (3.7) and (4.6).
[28] Benson, *op. cit.*, p. 127. Other examples include (4.5).
[29] Sanders, *op. cit.*, p. 340. For additional examples, see Cowgill, *op. cit.*, p. 152; Sabloff and Willey, *op. cit.*, p. 315; and Culbert, *Lost Civilization*, pp. 47–49. Perhaps (4.7) is an additional example.
[30] Sanders, *op. cit.*, p. 334; see also p. 362.
[31] Shimkin, *op. cit.*, pp. 269–299. Other examples include (2.05) and Saul, *op. cit.*
[32] Perhaps (2.13), (2.14), and (2.15) are negative arguments of type (2).
[33] For additional examples, see (4.3) and Cowgill, *op. cit.*, p. 152.
[34] Andrews, *op. cit.*, pp. 259–260. For additional examples, see (1.4), (2.08) and Cowgill, *op. cit.*, p. 152.
[35] Sanders, *op. cit.*, p. 364. For an additional example, see Thompson, *Rise and Fall of Maya Civilization*, p. 104.
[36] Sanders, *op. cit.* For additional examples, see Cowgill, *op. cit.*; and Adams, 'Maya Collapse', pp. 149–158.

PART III

OTHER TOPICS

R. M. MARTIN

FACT, FEELING, FAITH, AND FORM

Naśyami aham bhū naśyati lóka!
Sruyatâm dharma, Bhagawat.

[What is that, knowing which, we shall know everything?] It is not easy to know "the real, internal constitution" of God's nature, and perhaps no one has ever known it fully. Perhaps no one has ever known it even partially, although this is doubtful. Beliefs, intimations, surmises, and the like, have often sufficed. No matter, the notion of God should be characterized, it would seem, in so grand a fashion as to contain, in some specific sense, all knowledge of all beings and happenings, here, there, and everywhere, past, present, and future. In particular God's nature should contain, in a most intimate way, all scientific law, both causal and stochastic, as well as all boundary conditions. Hence implicitly it should contain all factually true statements. But God is not merely the repository of truth, but of value, of beauty, and goodness, as well. Science and value, whatever the shortcomings or defects of our knowledge about them, should be properly fused, it would seem, in any satisfactory characterization of the real internal constitution of God's nature. Failure to attain this fusion is to rest content with only a partial and hence an inadequate characterization.

The perennial theme "that being is one and identical with God the creator", as Richard Taylor puts it, ". . . is rediscovered in every age and in every corner of the world. It is at once terrifying and completely fulfilling. It will never perish and nothing will ever replace it. Nothing possibly can: its endurance is that of the stars".[1] But even the stars may come and go and still be terrifying. Only if we add the insight of *philosophia perennis*, that being is in its real nature *akin* to *mind* or *spirit*, in some sense, do we have the basis for a view of the kind described. Whatever spirit is, being is "identical" with it, and being *one*, so also is spirit. "There is only one river, which here and there assumes new forms or is modified in this way and that, either briefly or more lastingly. Here it assumes the form of a ripple, there of a waterfall, and numberless

Peter van Inwagen (ed.), Time and Cause, 239–254.
Copyright © 1980 *by D. Reidel Publishing Company.*

other forms in other places".[2] Being here and now is a material object, but there and then a mental act perhaps. No matter what forms or shapes it assumes or however it is modified, it still may be regarded as identical in character with God the creator, "that from which the origin, subsistence, and dissolution of this world proceed".

Let 'AS' be a primitive individual constant designating "absolute spirit" or "mind". Immediately we note, as a first metaphysical principle, that AS exists,

Pr1. $\vdash$E!AS.

The existence of individuals is handled here predicatively, where

'E!x' is short for '$\sim x = N$',

N being the null individual.[3]

It is interesting that Hegel, at the very beginning of his *Phänomenologie des Geistes*, differentiates "subjective" and "objective" spirit from the AS.[4] The one is a "manifestation" of AS, the other, we might say, is an "embodiment" of it. The various objects of nature are embodiments of AS; those of the mental realm, manifestations. Accordingly, two new primitives are needed for these notions. Let us symbolize them by 'Manif' and 'Emb'. Clearly the following principles should obtain concerning these notions.

Pr2. $\vdash (x)(y)((x \text{ Manif } y \vee x \text{ Emb } y) \supset x = \text{AS})$.
Pr3. $\vdash \sim(\text{E}x)(x \text{ Manif AS} \vee x \text{ Emb AS})$.

Thus AS alone manifests or embodies anything, and nothing whatsoever manifests or embodies it. Also nothing is both manifested and embodied by anything.

Pr4. $\vdash \sim(\text{E}x)(\text{E}y)(x \text{ Manif } y \cdot x \text{ Emb } y)$.

We may now define

'SubjSp' as 'Fu'$\{x \ni \text{AS Manif } x\}$'

and

'ObjSp' as 'Fu'$\{x \ni \text{AS Emb } x\}$'.

Thus the realm of subjective spirit is the fusion of (the virtual class of) everything manifested by AS, and objective spirit is the fusion of (the virtual class of) everything embodied by AS. These definitions give a very natural way of providing for the two Hegelian realms. Should they be regarded as mutually exclusive? If so, we need to postulate that every *part* of a manifested or embodied individual is also manifested or embodied, respectively. Thus, where P is the part-whole relation, we have also that

Pr5. $\vdash(x)(y)(z)((x \text{ Manif } y \cdot z \text{ P } y) \supset x \text{ Manif } z)$

and

Pr6. $\vdash(x)(y)(z)((x \text{ Emb } y \cdot z \text{ P } y) \supset x \text{ Emb } z)$.

Also it should then obtain that

$$\vdash\sim(\text{E}x)(\sim x = N \cdot x \text{ P SubjSp} \cdot x \text{ P ObjSp}),$$

that SubjSp and ObjSp have no nonnull part in common.

If these two spheres are taken to exhaust the cosmos, we have also a *principle of completeness*, that

Pr7. $\vdash(x)(\sim x = \text{AS} \supset (\text{AS Manif } x \vee \text{AS Emb } x))$.

Note of course that we are in no way following Hegel here but merely making use of some of his terminology.

But perhaps there are realms of derivative being other than these two, or even altogether different. Perhaps the two Hegelian ones are themselves unjustifiable on the basis of modern science, and constitute an illicit dichotomy. These difficult questions we need not attempt to answer for the moment, but we should note that the foregoing material may easily be extended to allow for any number of derivative realms of being – or even for none at all. But let us assume at least one. And let us speak of manifestation in a wider sense for the moment, so as to include embodiment, as well as whatever further kinds of process are appropriate for generating the given kinds of entities. Thus we let 'Manif_1', 'Manif_2', and so on, be primitives, and we let

'U_i' abbreviate '$\text{Fu}`\{x \ni \text{AS Manif}_i\, x\}$'.

Thus the universe of entities U_i is merely the fusion of the entities to which AS bears Manif_i, for each i. For each relation Manif_i we then have principles analogous to Pr2 and Pr3, and an appropriate extension of Pr4.

If $i = 0$, absolute monism results, AS is the only reality and there is nothing else except *māyā*. Even the name 'AS', the very inscriptions of Pr1–Pr4, and so on, would be dropped. They would all be items of *māyā* and thus presumably not worthy of rational discourse. But even if $t > 0$, we could still hold to a form of the doctrine of *māyā* in regarding the entities of U_1, U_2, and so on, as *māyā* items but allow rational discourse about them. However, if the discourse is to be in accord with modern logic and science, it will quickly be seen to be so important for our human life, and so insistently objective and compelling – and indeed so difficult to come anywhere near getting right – that the point of talk of *māyā* at all is soon lost. Surely the AS is not the less great, the less worthy of our total and all-absorbing effort to grasp it, if we regard the derivative entities to be genuine in some sense, if only as manifestations of it. In fact, the situation is the other way around. Let us embrace the derivative entities as worthy of our love and respect, and make every possible effort to come to see most intimately how they are interrelated one with another. It is in this way, in part, that we can come to know the grandeur and munificence of the AS itself. However, our "knowledge" of it need not be exhausted therewith, but rather enhanced.

The manifested objects of the U_i's are to comprise whatever it is that our cosmos contains. Precisely how we are to populate them is of course an incredibly difficult matter. Surely they must contain the objects needed for the sciences in their most developed stages. We must not rest content with the ontology of centuries back nor even with the "stale" science of yesterday. But to spell out in detail the ontology of even one science, at its present state of development, would be very difficult, and would tax even the greatest practitioners. Nonetheless, we may suppose it to consist of a presumably small number of U_i's in terms of which the desired assertions of that science can be made. And similarly for other sciences. And we must never suppose that any characterization of the U_i's needed for science would ever be final or complete. On the contrary, they would always be *semper reformanda*, and would exhibit enormous variation in the hands of different practitioners in the same field even at the same time.

Of particular interest for philosophers of logic and mathematics is the U_i, or the U_i's, needed for both. If logic is taken as standard, first-order logic, as throughout this paper, no assumption concerning the U_i's need be made. On the monist view, our only individual is AS, plus the null and world individuals. The latter, however, would be identical with AS, and the null individual N has the property that

$$\sim \mathrm{E!}N,$$

that it does not 'exist' in the appropriate sense. (Of course N is a value for a variable, but that is something else again.) And if $i > 0$, the U_i's are merely those of the sciences as already provided.

Logic as such has no ontology. For mathematics, however, the situation is very different. Let us think of it set-theoretically, in terms of the Zermelo-Fraenkel-Skolem system.[5] Here two U_i's are needed, one for individuals or

Urelemente – Zermelo himself insisted upon their admission, it will be recalled – and one for the realm of sets. No harm need arise from admitting the *Urelemente*, the very entities that may be presumed to populate the cosmos. The admission of a domain of sets, however, postulates entities that do not populate the cosmos in any obvious sense. Even so, this matter need not deter us, for we may use merely our *Urelemente* but allow set-theoretic talk about them in the manner of the "moderate" realism of Duns Scotus.[6] In this way classical mathematics in the set-theoretical sense may be preserved, and used, moreover, as a basis for the other theoretical sciences. For this, of course, a new primitive is needed, and for applications to the sciences, such new primitives as those sciences require.

The question arises as to whether, once the U_i's required for the sciences have been arrived at, any further ones are needed. Do the ontologies of the sciences suffice for all discourse – other, of course, than that concerning the AS and its possible manifestations in SubjSp? Well, surely yes, if 'science' is construed widely enough. Note that the question is merely one about *ontologies*, not about the modes of discourse allowed concerning the items admitted in that science. One and the same act, for example, may be said to occupy such and such a place-time in one context, but to be immoral or illegal or prohibited or whatever, in others.

Mental entities are the occupants of the realm of SubjSp, and any interesting metaphysical idealism may be presumed to admit such entities. The basic items here are no doubt individual souls or minds, and mental acts are presumably fundamentally dependent upon these. It is a bit mysterious as to just what an individual mind is and how it is to be individuated. You have one and I have one, and they are alike in both being minds. Let yours be m_1, and mine m_2, and let m_3, m_4, and so on, be those of others. The calculus of individuals allows us to form, then, the "group" mind

$$(m_1 \cup m_2 \cup m_3 \cup m_4 \cup \ldots).$$

Even this group mind does not of course exhaust the AS, the latter being infinitely greater. Is this group mind a *part* of the AS? If so, then each individual mind is also, each being a part of the group sum. Equally difficult is the question as to how the individual souls or selves are related to the mental acts of or pertaining to them. And this in turn leads to the problem as to how such acts themselves are to be individuated.

Individual minds result from the AS by one kind of manifestation, bodies by another. Does the human person, a unique complex of mind and body, result by still a third kind of manifestation? Some idealists might well contend so. To bring then a mind, a body, and a person together, we need the Of-relation of possession.[7] It is not clear whether the mind possesses the body, or the body the mind, or the person the mind, or the mind the person, or the body the person, or the person the body. Perhaps there is possession in all of these ways. In any case, if bodies, minds, and persons result from separate kinds of manifestation, a suitable way of bringing a body, a mind, and a person together must be at hand to provide for a concrete human person.

The idealist, of course, regards minds as *par excellence* the real entities, they being like unto the AS itself. Rather than regard the other types of entities as arising by other kinds of manifestation, perhaps they should be regarded rather as the result of the *concentration* of soul-stuff in some particular way or other. Each material object is merely soul *concentrated* in a certain way. The notion of concentration, the very prototype of mental activity, then would play the role of the relations of manifestation. But concentration is mental in a way in which the relations of manifestation are not. And if "subject" and "object" are alike, both must be mental. Thus the following "principle", where "Conc" is the relation of concentration, might well hold, namely,

$$\vdash (x)(\text{AS Conc } x \supset x \text{ Like AS}).$$

Everything that results from the AS by concentration is itself like or similar to AS. And likewise,

$$\vdash (x)(x \text{ Like AS} \supset x \text{ P AS}),$$

that everything like AS is itself a *part* of it. The former principle might well hold without the latter. If the two principles are taken together, a genuine monism, even a pantheism, is achieved. The development of idealism in terms of the theory of concentration would be more Vedantic than Hegelian. Principles akin to Pr2–Pr4 and Pr6–Pr7 would obtain, with 'Conc' in place of 'Emb', no change being required in Pr1 and Pr5.

For the purposes of the subsequent discussion, and to simplify, let us presuppose the theory above as developed in terms of the Hegelian 'Manif' and 'Emb'. But whatever modifications of this might be thought desirable can easily be presupposed equally well.

However the fundamental ontology is arranged, the AS has remarkable tasks to perform and must be given some remarkable properties, akin to those of the Thomistic God and the Whiteheadian primordial nature. To see this let us consider again the primordial valuations constituting this latter, which will be helpful as a heuristic, enabling us to flesh out the theory underlying St. Thomas's "five signs of will".

The five signs of will, it will be recalled, are *operation* (doing), *permission, precept, counsel,* and *prohibition,* but St. Thomas is not too clear as to precisely how these are to be construed. The words are used analogically. "A man may show that he wills something . . ." by doing it

> directly when he works in his own person; in that way the expression of his will is his own *working*. He shows it indirectly, by not hindering the doing of a thing: . . . In this respect the expression is called *permission*. He declares his will be means of another when he orders another to perform a work, either by insisting upon it as necessary by *precept*, and by *prohibiting* its contrary; or by persuasion, which is a part of *counsel.*[8]

St. Thomas goes on to note that

> since the will of man makes itself known in these ways, the same five are denominated with regard to the divine will, as the expression of that will. That *percept, counsel,* and *prohibition* are called the will of God is clear from the words of *Matth*. vi. 10: *Thy will be done on earth as it is in heaven*. That *permission* and *operation*,are called the will of God is clear from Augustine, . . . who says: *Nothing is done, unless the Almighty wills it to be done, either by permitting it, or by actually doing it.*[9]

These very significant but difficult comments should be helpful in attempting to characterize the divine will, whether construed Thomistically or not.

Among the operations we should surely include all the manifestations and embodyings. These operations concern only the ontology. In addition, there are the primordially ordained circumstances, laws, and so on. Let

'AS PrimOp $a,x_1, \ldots, x_n$'

express that the AS *primordially* operates or has it obtain that the *n*-place predicate *a*, standing for a virtual class or relation, apply to or denote $x_1, \ldots,$ x_n, in this order.

That the AS is the "creator" of all entities (other than himself) is in effect provided by Pr7 above. But he is also the ordainer of all scientific, moral, and aesthetic law, and this aspect of the divine activity can be stipulated only by bringing in the relation PrimOp. Thus suppose $\ulcorner a,x_1 \ldots x_n \urcorner$ obtains, for fixed $x_1, \ldots, x_n$, and *a*, not just factually but as the result of, or as an instance of, some scientific law. Then it would obtain that

Pr8. $\vdash$AS PrimOp $a,x_1, \ldots, x_n$,

for such $a,x_1, \ldots, x_n$. Nor need Pr8 be restricted to just scientific law. It should be extended to instances of whatever laws are thought to obtain in any of the spheres of knowledge. And if one or more of the x_i's are allowed to be numbers, natural, real, or complex, even laws of a probabilistic kind may also be included here. Think what a staggering principle Pr8 then is, incorporating as it does all the laws governing the cosmos, construed in the most inclusive possible sense. But surely the AS must be conceived as so great as to incorporate no less.[10] Clearly also it holds that

Pr9. $\vdash (y)(a)(x_1) \ldots (x_n)(y \text{ PrimOp } a, x_1, \ldots, x_n \supset y = \text{AS})$,

so that the AS is the *only* entity capable of the primordial operations. And also

Pr10. $\vdash (a)(x_1) \ldots (x_n)((\text{AS PrimOp } a, x_1, \ldots, x_n \cdot a \text{ Des}_{\text{VC}} F) \supset Fx_1, \ldots, x_n)$.

Whatever is primordially ordained to obtain does actually obtain. But the converse need not hold. Not all that obtains is primordially ordained to do so. ('Des_{VC}' is the sign for the designation of virtual classes or relations, and if $n = 1$

'$a \text{ Des}_{\text{VC}} F$' is short for '$(\text{PredCon}_1\, a \cdot (x)(a \text{ Den } x \equiv Fx))$',

'Den' being the primitive for denotation.[11])

The primordial operations need not be confined to just the demands of scientific law, as already noted. Moral and aesthetic laws, if there are such, are included, and even such boundary conditions as might be thought to obtain independent of law. Perhaps even there are miracles in some sense as the direct result of a primordial operation. If so, the stipulation of such is presumed included here.

St. Thomas speaks of prohibition in a somewhat narrow sense, of prohibiting the "contrary" of a precept. Here let us speak rather of prohibiting the contradictory of an operation. Thus we may let

'$\text{AS PrimPrhbt}_{\text{Op}}\, a, x_1, \ldots, x_n$' abbreviate '$\text{AS PrimOp} \ulcorner -a \urcorner, x_1, \ldots, x_n$',

where $\ulcorner -a \urcorner$ is the negation of a. There are other kinds of prohibition, which we shall meet with in a moment.

The primordial operations concern all objects whatsoever, including human persons, actions, events, states, processes, and the like. The precepts and counsels, on the other hand, may be thought to concern only human beings and their actions. Let 'p' be a variable for persons and 'e' for actions of the kind humans are capable of performing. And let P be a virtual class of persons satisfying such and such conditions, and A a class of suitable actions. Then we may let

'AS PrimPrcpt 'P', '$\{p \ni (\mathrm{E}e)(p \text{ Prfm } e \cdot Ae)\}$' '

express that it is a primordial precept that persons of the kind P should be persons who *perform* actions of the kind A, under appropriate circumstances. Precepts always seem to be general in this way applying to all persons and actions of given kinds. Counsels, on the other hand, may always be regarded as specific, applying to a given person with respect to a given action.

Are all counsels covered by a precept? It is tempting to think so, whether the precept is explicitly known or exhibited or not. If so, we may let

'AS PrimCnsl p,e, 'P', 'A' ' abbreviate '(AS PrimPrcpt 'P', '$\{q \ni (\mathrm{E}e')(q \text{ Prfm } e' \cdot Ae')\}$' $\cdot Pp \cdot Ae$)',

so that p is counseled to do e relative to P and A just where it is precepted that all P's do A's and p is a P and e an A.

More general definitions, with variables in place of the constants, may be given by letting

'AS PrimPrcpt a,$\ulcorner\{p \ni (\mathrm{E}e)(p \text{ Prfm } e \cdot b \text{ Den } e)\}\urcorner$'

and the like also be significant primitively, and then letting

'AS PrimCnsl p,e,a,b' abbreviate '(AS PrimPrcpt a,$\ulcorner\{q \ni (\mathrm{E}e')(q \text{ Prfm } e' \cdot b \text{ Den } e')\}\urcorner \cdot a \text{ Den } p \cdot b \text{ Den } e$)'.

Note that by means of precept the AS in effect "orders" a person "to perform a work" by "insisting upon it as necessary", in some social, moral, or aesthetic sense. And surely some generality must obtain as a condition for the necessity. Hence the use of the class terms 'P' and 'A'. Counsel, however, is always specific and 'persuasion' is a part of it. Only a person, even a sum of persons, can be persuaded and hence counseled in this sense.

There are relevant kinds of prohibition corresponding to precept and counsel. Thus we let

'AS $\text{PrimPrhbt}_{\text{Prcpt}}$ 'P', 'A' ' abbreviate 'AS PrimPrcpt 'P','$\{p \ni {\sim}(\mathrm{E}e)(p \text{ Prfm } e \cdot Ae)\}$' ',

so that persons of the kind P are prohibited in this sense from being persons who perform actions of the kind A. And there are also prohibitive counsels, so that

'AS $\text{PrimPrhbt}_{\text{Cnsl}}$ p,e,'P','A' ' abbreviates '(AS $\text{PrimPrhbt}_{\text{Prcpt}}$ 'P','A' $\cdot Pp \cdot Ae$)'.

More general forms of these definitions, with variables in place of the constants 'P' and 'A', may also be given.

Clearly, corresponding with Pr9, we should have that

Pr11. $\vdash (x)(a)(b)(x \text{ PrimPrcpt } a,b \supset (x = \text{AS} \cdot (y)(a \text{ Den } y \supset \text{Per } y) \cdot (y)(b \text{ Den } y \supset \text{Per } y)))$,

when 'Per' is the predicate for persons.

Also where

'p Oblg a'

expresses deontically that p is *obliged* to be a person of the kind denoted by a, we should have that

Pr12. $\vdash (a)(b)(p)((\text{AS PrimPrcpt } a,b \cdot a \text{ Den } p) \supset p \text{ Oblg } b)$.

This principle assumes that whatever is primordially precepted, so to speak, is deontically obliged. This at least should hold, but not the converse. There are surely obligatory acts not determined so primordially.

No doubt much takes place in the cosmos that is primordially neutral, in the sense of being neither the result of an operation nor operationally prohibited. Thus, where 'PredCon_n a' express that a is an n-place predicate constant,

'AS $\text{PrimNtrl}_{\text{Op}}$ $a,x_1, \ldots, x_n$' may abbreviate '(PredCon_n a $\cdot$ $\sim$AS PrimOp $a,x_1, \ldots, x_n$ $\cdot$ $\sim$AS $\text{PrimPrhbt}_{\text{Op}}$ $a,x_1, \ldots, x_n$)'.

And similarly for human actions that are neither covered by precept nor prohibited by precept. Thus also

'AS $\text{PrimNtrl}_{\text{Prcpt}}$ 'P','A' ' abbreviates '($\sim$AS PrimPrcpt 'P','A' $\cdot$ $\sim$AS $\text{PrimPrhbt}_{\text{Prcpt}}$ 'P','A')'.

Here too, a more general definition may easily be given.

Note that in the foregoing only 'PrimOp' and 'PrimPrcpt' have been needed as primitives, in addition of course to 'Manif', 'Emb', and 'AS'. All the other primordial predicates have been defined within the linguistic framework embodying quantification theory, identity, mereology (or the calculus of individuals), and of course some semantics and event theory. The deontic notion 'Oblg' is also presumed available, either primitively or by definition, but it is not a purely primordial notion, being relative always to a given social group and a specific deontic code.

There is also the all-important notion of a primordial *permission*, to which we now turn.

It is clear, if nothing is done other than its being done either by the Almighty or being permitted by him, that the operations and permissions exhaust the divine will and that the other "signs" are to be handled as subdivisions. The operations and permissions are thus to be mutually exclusive and jointly exhaustive. Under the primordial operations are included the manifestations, embodiments, operations proper, and the operational prohibitions. These operations are all such that their results, so to speak, *must* obtain if our cosmos is to be the way it is. All the other primordial notions are included in the permissions, whose results *may* be violated in our cosmos. Note the implicit distinction here between the operations and permissions, on the one hand, and their "results", on the other. The results of the one *must* obtain, but those of the other need not. On the other hand, the operations and permissions themselves constitute the necessary activity of the AS, if our cosmos is to be what it is.

The prohibitions include just the three kinds, operational, preceptual, and counsel-wise, the precepts both the proper and prohibitive ones, and similarly for the counsels. The primordially neutral comprise the operationally neutral and the preceptually so. The primordial permissions, as already noted, then comprise all the primordial activities not included in the operations, i.e., the prohibitions, the precepts, the counsels, and the primordially neutral. These comments may all be summarized by means of three additional definitions. We may let

'PrimOp e' abbreviate '(Ea)(Ex_1) . . . (Ex_k) (⟨AS,Manif,x_1⟩e v ⟨AS,Emb,x_1⟩e v ⟨AS,PrimOp,a,x_1⟩e v ⟨AS,PrimOp,a,x_1,x_2⟩e v . . . v ⟨AS,PrimOp,a,x_1, . . . , x_k⟩e)',

'PrimPrmsn e' abbreviate '(Ea)(Eb)(Ep)(Ee')(Ex_1) . . . (Ex_k) (⟨AS,PrimPrhbt$_{\text{Prcpt}}$,a,b⟩e v ⟨AS,PrimPrhbt$_{\text{Cnsl}}$,p,e',a,b⟩e v ⟨AS, PrimPrcpt,a,b⟩e v ⟨AS,PrimCnsl,p,e',a,b⟩e v ⟨AS,PrimNtrl$_{\text{Op}}$,a,x_1⟩ e v . . . v ⟨AS,PrimNtrl$_{\text{Op}}$,a,x_1, . . . , x_k⟩e v ⟨AS,PrimNtrl$_{\text{Prcpt}}$, a,b⟩e)',

and

'PrimPrhbtn e' abbreviate '(Ea)(Eb)(Ep)(Ee')(Ex_1) . . . (Ex_k) (⟨AS,PrimPrhbt$_{\text{Op}}$,a,x_1⟩e v . . . v ⟨AS,PrimPrhbt$_{\text{Op}}$,a,x_1, . . . , x_k⟩e v ⟨AS,PrimPrhbt$_{\text{Prcpt}}$,a,b⟩e v ⟨AS,PrimPrhbt$_{\text{Cnsl}}$,p,e',a,b⟩e)'.

These definitions introduce the notions of being a primordial operation, permission, or prohibition, respectively.

Note the use of the variable 'e' for an act or state. And recall that the expressions enclosed in the angle brackets are event-descriptive predicates. Thus '$\langle \mathrm{AS},\mathrm{Manif},x_1\rangle e$', for example, expresses that e is an act or state of x_1's being manifest by AS. Extensive use is made of such predicates within event logic.[12] Recall also the special use of the parameter 'k' for the degree of the primitive predicate of greatest degree needed as a primitive, and where there are assumed to be primitive predicates of each degree n where $1 \leqq n \leqq k$.

The notion of the divine will may be thought to be fully analyzed in terms of the disjunction of these three. Thus

> 'DW' may be short for '$\{e \ni (\mathrm{PrimOp}\ e \vee \mathrm{PrimPrmsn}\ e \vee \mathrm{PrimPrhbtn}\ e)\}$'.

The DW is thus merely the virtual class of all primordial operations, permissions, and prohibitions.

A few principles over and above Pr1–Pr2 that should presumably obtain are as follows.

Pr13. $\vdash (a)(x_1) \ldots (x_n)(\mathrm{AS\ PrimOp}\ a,x_1,\ldots,x_n \supset (\mathrm{PredCon}_n\ a \cdot {\sim}x_1 = \mathrm{AS}, \ldots . {\sim}x_n = \mathrm{AS}))$.

Pr14. $\vdash {\sim}(\mathrm{E}a)(\mathrm{E}x_1) \ldots (\mathrm{E}x_n)(\mathrm{AS\ PrimOp}\ a,x_1,\ldots,x_n \cdot \mathrm{AS\ PrimPrhbt_{Op}}\ a,x_1,\ldots,x_n)$.

Pr15. $\vdash {\sim}(\mathrm{E}a)(\mathrm{E}b)(\mathrm{AS\ PrimPrcpt}\ a,b \cdot \mathrm{AS\ PrimPrhbt_{Prcpt}}\ a,b)$.

Pr16. $\vdash (a)(b)(x_1) \ldots (x_n)(\mathrm{AS\ PrimOp}\ \ulcorner(a \cup b)\urcorner,x_1,\ldots,x_n \equiv (\mathrm{AS\ PrimOp}\ a,x_1,\ldots,x_n \cdot \mathrm{AS\ PrimOp}\ b,x_1,\ldots,x_n))$,

Pr17. $\vdash (a)(b)(x_1) \ldots (x_n)((\mathrm{AS\ PrimOp}\ a,x_1,\ldots,x_n \vee \mathrm{AS\ PrimOP}\ b,x_1,\ldots,x_n) \supset \mathrm{AS\ PrimOp}\ \ulcorner(a \cup b)\urcorner,x_1,\ldots,x_n)$.

Pr18. $\vdash (a)(b)(x_1) \ldots (x_n)((\mathrm{AS\ PrimOp}\ a,x_1,\ldots,x_n \cdot \mathrm{AS\ PrimOp}\ \ulcorner(-a \cup b)\urcorner,x_1,\ldots,x_n) \supset \mathrm{AS\ PrimOp}\ b,x_1,\ldots,x_n)$.[13]

Some of the various principles given may need some modification in the light of a more thorough presentation. The whole theory of primordiality in fact cries out for further elaboration and development, being still in its infancy.[14]

The analysis of the primordial valuations and hence of the divine will, given above, agrees with that of St. Thomas to some extent. A few additional points of parallel are as follows. St. Thomas notes that "there is no reason why the same thing should not be the subject of precept, operation, and counsel; or of prohibition or permission".[15] Clearly one and the same human act can be the result of a prohibitional operation as well as a prohibitional counsel, and hence of a prohibitional permission. St. Thomas contends also that "God ordains rational creatures to act voluntarily and of themselves. Other creatures act only as moved by the divine operation; therefore only operation and permission are concerned with these".[16] This contention agrees with the foregoing, permission here being taken in the sense of the primordially neutral.

"All evil of sin", St. Thomas notes also, "though happening in many ways, agrees in being out of harmony with the divine will. Hence with regard to evil, only one expression [of will] is assigned, that of prohibition."[17] The evil of sin is precisely what is primordially prohibited by precept. (There is no sin as the result of a primordial operation, all such being in the primordially good.) "On the other hand", St. Thomas goes on, "good [the humanly good] stands in various relations to the divine goodness, since there are good deeds without which we cannot attain to the fruition of that goodness, and *these are the subject of precept*."[18] The primordially good is the subject of precept, and counsel above was taken as instantial of precept. But St. Thomas construes counsel here rather differently, "and there are others [goods]", he says, "by which we attain to it [the fruition] more perfectly, and these are the subject of counsel".[19] Here counsel seems to be concerned rather with supererogation. But even some precepts might be stipulative of the supererogatorily good, so that even this last remark could be seen to accord with the foregoing.

Any philosophical discussion of God's will must perforce be speculative, as indeed is the foregoing. There would not seem to be much point in discussing it at all, however, without some analysis of what the phrase is supposed to designate. At best we can merely hypothesize what this might be, and thus we never could be said to know it in any more direct sense. Even so, hypothetical constructs are useful in theology just as they are in theoretical science.[20]

Note that the foregoing hypothetical reconstruction of some features of metaphysical idealism has been given in a semantical metalanguage incorporating a theory of acts. It would seem very doubtful that a more restricted kind of logical framework would suffice for this purpose. Note also that the primordial notions have been handled intensionally. These are given by reference to a *predicate* rather than to a (virtual) class or relation the predicate might designate. The reason for this is the familiar one concerning the intensionality of obligation, to which the primordial notions are akin. It would not do to say, in a deontic logic, for example, that one is obliged to be an F, for F might be equivalent to some G, with respect to which one is not obliged. Reference to the predicate 'F' here instead of to the virtual class F prevents any such unwanted consequence. Hence the intensional treatment, within a semantical metalanguage, of the primordial notions throughout, in terms essentially of Frege's *Art des Gegebenseins*.

An alternative, more sophisticated way of handling manifestation and embodiment, and even some of the primordial relations, suggests itself if a numerical measure is introduced. We may think of the AS as manifesting itself in *x to just such and such a degree*. All entities manifested to the same degree would then be of essentially some same kind. The very difference between manifestation and embodiment could then be handled in terms of difference of degree. Embodiment would be low degree of manifestation. Let

'AS $\mathrm{Manif}^i\, x$'

express that x is a manifestation of AS to just degree i. If $i = 0$, we could let x be the null-entity, and if $i = 1$, we could let x be AS itself. AS then manifests itself to maximal degree. Physical objects have low degrees attached to them, and highly mental ones have high degrees. And similarly for the primordial precepts, some of which are more binding than others. Here too it might be of interest to introduce a numerical degree. Whitehead speaks of the degree of a primordial valuation, as noted above. No one, it would seem, has ever developed such a theory in any detail, however, for natural theology and the use of numerical measures are not ordinarily thought to go hand in hand. A quite sophisticated view would result if a suitable numerical measure were introduced, and no doubt some interesting notions would be forthcoming in terms of it.

Nothing has been said thus far concerning physical time, space, causality, and the like. Any attempt to locate the AS with respect to any of these is quite foreign to the foregoing. It is rather the other way around, all objects of the physical world themselves being embodiments of the AS. Hence the foregoing theory is all couched in the Fregean tense of timelessness, so to speak, as in that of spacelessness, causalitylessness, and so on.

Of course, only the barest logical *maquette* of the full theory concerning AS has been given here. Indeed, to flesh out the foregoing in adequate detail would be a formidable task indeed. Nonetheless, certain general features of what the fuller development would be like should be evident. In particular it would comprise foundations for a theory of objective value as contained in the primordial precepts. Thus, as far as this scheme goes, there is no essential dichotomy between fact and value, but each is handled in its separate way. Nor is there any easy reduction of one to the other. Each is given its proper dignity and the way is left open for discriminating all manner of interconnections between the two. Note also that there is here no illicit dichotomy between reason and faith. Again, it is rather that a rational scheme is available in which a theory of faith may be incorporated. Indeed, it may be that faith, in a suitable sense, is our highest rational activity, for it is always reasonable to let one's mind wander to an *O altitude*! The task of natural theology in fact may be thought to be just this.

But faith is nothing if it does not issue in action, as many writers in the tradition of *philosophia perennis* have eloquently affirmed. And indeed the notion of the AS is of such staggering grandeur and magnitude, that it seems eminently rational that we should "shape the whole conduct of [our] life ... into conformity" with it. To do this, in fact, should be our whole aim, everywhere and always, as the great writers of that tradition have been continually affirming across the centuries. "To interpret the absolute we must give all our time to it." The pursuit of science, of beauty, and of goodness are alike here given their proper role in this endeavor.

There is something compelling about human feeling at what we take to be its highest, in the full experience, say, of a great work of art. It is doubtful that such feeling can be suitably and fully explicated on any other basis than one such as the foregoing. We can go a long way in analytic aesthetics without it, but always with a most essential human ingredient left out – the depth and quality of authentic aesthetic feeling as its best.

The positive contribution of the present paper is merely to have made some tentative suggestions toward giving the *philosophia perennis* the logical backbone it is often thought to lack. Usually in discussions of the AS there is too much logically irresponsible misstatement. But so lofty a topic would seem best served by using such clean-cut logical notions as are now available. Surely we should let idealism, along with other metaphysical views, grow with the advance of science and philosophic technique.

Northwestern University

NOTES

1 Richard Taylor, *With Heart and Mind*, St. Martin's, New York, 1973, Proem.

2 *Ibid.*, p. 142.

3 On the null individual, see especially the author's 'Of Time and the Null Individual', *Journal of Philosophy* **62** (1965), 723–736. To characterize the null entity we need of course the calculus of individuals.

4 G. W. F. Hegel, *Werke*, Vol. 2: *Phänomenologie des Geistes*, Duncker and Humblot, Berlin, 1841.

[5] For useful expository remarks, see especially H. Wang, *From Mathematics to Philosophy*, Humanities Press, New York, 1974.

[6] See the author's 'On Common Natures and Mathematical Scotism', *Ratio* **19** (1977), 103–114. Reprinted in R. M. Martin, in *Peirce's Logic of Relations and Other Studies*, Peter de Ridder, Lisse, 1979, pp. 136–147.

[7] Cf. the author's 'Of "Of"', in his *Pragmatics, Truth, and Language*, D. Reidel Publishing Co., Dordrecht, 1979, Chap. X.

[8] St. Thomas Aquinas, *Summa Theologica* 1.19.12, Burns Oates and Washbourne, London, 1920, Vol. 1, pp. 283–284, emphasis added.

[9] *Ibid.*, p. 284, some emphasis added.

[10] P8 is of course oversimplified, but a more general formulation and discussion is not needed for the present.

[11] *Cf.* the author's *Truth and Denotation*, University of Chicago Press, Chicago, 1958, p. 106.

[12] See the author's *Events, Reference, and Logical Form*, Catholic University of America Press, Washington, D. C., 1978; and *Semiotics and Linguistic Structure*, State University of New York Press, Albany, 1978.

[13] The '$\cup$' and '$\cap$' are the signs for the union and intersection respectively of virtual classes or relations.

[14] *Cf.* an alternative treatment of the primordial valuations in the author's 'On God and Primordiality', *Review of Metaphysics* **29** (1976), 497–522. See also 'Some Thomistic Properties of Primordiality', *Notre Dame Journal of Formal Logic* **18** (1977), 567–582; and 'The Logic of Idealism and the Neglected Argument', in Martin, *Peirce's Logic*, pp. 110–120.

[15] *Loc. cit.*

[16] *Ibid.*, p. 285.

[17] *Ibid.*

[18] *Ibid.*, emphasis added.

[19] *Ibid.*

[20] *Cf.* J. Bochénski, *The Logic of Religion*, New York University Press, New York, 1965. See also the discussion thereof in Martin, *Whitehead's Categoreal Scheme and Other Papers*, Martinus Nijhoff, The Hague, 1974, Chap. 9.

CORRIGENDA

p. 248, l. 8 f.b., for the right-most 'A' please read '$\{p \ni (\mathrm{E}e)(p \ \mathrm{Prfm}\ e \cdot Ae)\}$'.

p. 250, in Pr15, for 'AS $\mathrm{PrimPrhbt}_{\mathrm{Prcpt}}$ a, b' please read 'AS PrimPrcpt a,⌜$-b$⌝'.

JOEL FEINBERG

ABSURD SELF-FULFILLMENT

An Essay on the Merciful Perversity of the Gods

Richard Taylor adds a twist to the ancient legend of Sisyphus, who was condemned by the gods to perpetual life spent pushing a large rock to the top of a hill from which it falls down the other side, once more to be pushed to the top, and so on forever. "Let us suppose", writes Taylor, "that the gods, while condemning Sisyphus to the fate just described, at the same time, as an afterthought, waxed perversely merciful by implanting in him a strange and irrational impulse, namely a compulsive impulse to roll stones ... e.g. through implanting some substance that has this effect on his character and drives".[1] Such a modification would be merciful but also 'perverse', Taylor maintains, because "from our point of view there is clearly no reason why anyone should have a persistent and insatiable desire to do something as pointless as that".[2]

Taylor's Sisyphus, meaningless as his activities may seem to us, at least finds his rock-pushing career *fulfilling.* Since a powerful disposition to push rocks has been built into him, he is only doing what he is inclined by his nature to do when he pushes the rock, just as a dog fulfills his nature by chasing a rabbit or a bird by building a nest. One can criticize Taylor, incidentally, for his apparent confusion of self-fulfillment (doing what it is in one's nature to do) with compulsion. In Taylor's example, a substance in Sisyphus's blood forces him to 'want' to push stones, just as repeated injections of heroin into the veins of an unwilling prisoner would impose an addiction for heroin on him and make him 'want' his subsequent 'fixes'. The causal mechanism employed by the gods, however, need not be that crude, and we can imagine that they remake Sisyphus' nature in a more thoroughgoing way so that the disposition to push large objects, stemming from a reconstructed complex of glands and nerves and basic drives, becomes an integral part of Sisyphus' self rather than an alien element restraining him. Let us add a twist to Taylor's twist, then, and have the gods provide Sisyphus with a new nature rather than imposing an addiction on his old one. We can think of a rock-pushing Sisyphus as no more 'addicted' to his characteristic activities than we are to walking upright or to speaking a language. Our Sisyphus' activities, furthermore, are self-fulfilling, not simply because they satisfy his desires, nor because they involve his own wilful acquiescence, but simply because they express some basic genetic disposition of his nature.

Peter van Inwagen (ed.), Time and Cause, 255–281.
Copyright © 1980 *by D. Reidel Publishing Company.*

Taylor does not use the word ***absurd*** in describing Sisyphus' peculiar activities, but a whole tradition, whose most recent prominent member was Albert Camus[3], finds that term eminently appropriate. The words Taylor uses are "meaningless", "pointless", and "endless". Perhaps endless repetitive cycles of 'pointless' labor to no apparent purpose or result is only one species of absurdity, or perhaps 'pointlessness' is only one among many grounds for judging an activity to be absurd. In either case pointlessness and generic absurdity are not identical notions. But few would deny the synthetic judgment that there is an absurdity in pointless labors that come to nothing. Taylor isolates this absurdity by contrasting it both with painfulness and loneliness. It is not because Sisyphus' labors are arduous and body-bruising that they are absurd, for we can imagine that his rock is small and his labors undemanding. They would be no less pointless, and therefore no less absurd for that. Moreover, as Taylor suggests, we could give Sisyphus some partners so that the rock-moving activities are conducted by teams of two or more persons. That would reduce the loneliness of the enterprise but not its silliness. The essential absurdity of pointless activity is captured in a non-Sisyphean example that Taylor provides himself: "Two groups of prisoners, one of them engaged in digging a prodigious hole in the ground that is no sooner finished than it is filled in again by the other group, the latter then digging a new hole that is at once filled in by the first group and so on endlessly".[4]

Many philosophers have said that insofar as human existence is absurd there is a ground for certain negative attitudes – suicidal despair, philosophical pessimism, Camus' 'scorn', or defiance. Other philosophers, addressing a somewhat different datum, have said that insofar as a given human life is self-fulfilled it is a good life, and provides a reason for certain positive attitudes toward the human condition – hope, satisfaction, acceptance, or reconciliation. Often the 'optimists' say that some lives (at least) are completely fulfilled and most lives fulfilled to some degree or other. There is no necessity that all be fulfilled or that all be unfulfilled; it all depends on luck and skill. 'Pessimists', on the other hand, claim that all lives are necessarily absurd (meaningless, pointless), so their view is more sweeping. In any event, 'absurdity' and 'self-fulfillment' are quite different notions, so the 'optimists' and the 'pessimists' are not even talking about the same thing. Taylor suggests, quite plausibly, that life might be *both* absurd and (at its best, sometimes) self-fulfilling. What are we to make of that combination of truths? What are its consequences for optimism and pessimism? What general attitudes are appropriate if it is accepted? These questions call for closer examination of the concepts of 'absurdity' and 'self-fulfillment' and how they might fit together, and some comments on the question of how we can judge the appropriateness of cosmic attitudes.

I. ABSURDITY

One element that the various sorts of absurd things have in common is their irrationality, whether that be the apparently knowing assertion of manifestly false propositions or the apparently voluntary making of manifestly unreasonable decisions, or the apparently eager living of a manifestly pointless life.

A second element in all absurdity is implicit in the first, but deserves to be clearly stated on its own. Where there is absurdity there are always two things in clash or disharmony, distinguishable entities that conflict with one another. This element is referred to variously as the 'divorce',[5] disparity, discrepancy, disproportion, or incongruity between discordant objects. In general, things that do not *fit* together – means discrepant with ends, premises incongruous with conclusions, ideals disharmonious with practice, pretensions in conflict with realities – are found wherever there is absurdity. But having located the absurdity, we may attribute it either to the relation of disharmony itself, or exclusively to one, or the other, of the discordant objects.

In D. H. Parker's aesthetic theory,[6] the comic as well as the absurd thrives on discrepancy. One of the two discordant entities in comedy is always a *standard* and the other an object or performance that fails somehow to measure up to the standard. We perceive an object as 'pretending but failing to fit' some norm and then take pleasure in the incongruity. In some cases we adopt the point of view of the standard and 'laugh down' at the incongruous object, as when we delight in the undignified fall of the pompously powerful. In other cases we take the opposite viewpoint, that of the comically discrepant object itself, and we laugh at the standard, as for example when we laugh at cute children masquerading as adults, or in a quite different example, we laugh at a risqué story and thus have some fun with the sexual conventions violated in the tale. Perhaps not all funny things are absurd, and surely not all absurd things are funny, but discordance is an element common to comic and absurd things.

Another form of disparity is described in Thomas Nagel's astute account of absurdity,[7] namely the clash or disharmony between various *perspectives* from which we form attitudes and make judgments. There is an unavoidable discrepancy between the natural subjective way of viewing ourselves – as precious in our own eyes, full of genuinely important projects, whole universes in ourselves, persons who 'live only once' and have to make the most of the time allotted them – and various hypothetical judgments made from a more universal perspective – we are mere specks or drops in the ocean, or one of the teeming hive, absolutely inessential to the grand scheme of things, no more lovable in ourselves then are any of the zillions of individual insects whose infinitesimally transitory lives are equally as unimportant in the long run as our own. Our subjective point of view is an expression of the 'seriousness' with which all living beings must view their situations, a necessary expression of our biological natures. But the broader perspectives that yield a different and conflicting picture are available to any being with imagination and minimal conceptual development. And judged from those perspectives, the human expenditures of effort and emotion in the pursuit of 'important goals' are just so many posturings, and we mortals are absurd figures who strut and fret our hour upon the stage. The absurdity in the human condition, Nagel tells us, consists in a special kind of conspicuous discrepency, that between unavoidable pretension or aspiration, on the one hand, and reality as perceived from a truer perspective, on the other.

A word should be added here about the conceptual relation between the absurd and the comic. One of the most prominent classes of objects of our laughter is that of the absurdly incongruous.[8] Since not all absurdities are funny, however, a philosophical theory must explain what makes some incongruity comic and some not, and also why it is that we should take such pleasure in the incongruous when we do find it funny – why the sudden perception of absurdity in a situation can trigger the biological laugh reflex and produce pleasure in the laughing person. These philosophical problems remain unsolved, but of the standard theories of the comic, that of Max Eastman seems to me to be the most enlightening. We find nothing comic, Eastman reminds us, unless we are in at least a moderately playful mood. Sisyphus will find nothing remotely amusing in the absurdity of his situation if he is (as Camus would have him be) dead serious about it. An outsider, on the other hand, can find amusement in another's plight in proportion to the absurdity he finds in it, if he detaches the absurdity, so to speak, from the serious elements in the situation, and responds to it playfully.

Eastman points out that there are two effective ways of amusing a small child, once you have put him at his ease and restored him to his natural condition of playfulness: you can "make a perfectly terrible face", or you can try a kind of gentle practical joke on him, teasing him in a lighthearted spirit. These two techniques, Eastman points out, correspond to the two most famous definitions of laughter. Corresponding to the frightful face made playfully is Aristotle's definition of the comic as "some defect or ugliness which is not painful or destructive". All perceptual comic objects (what Eastman calls "ludicrousness") fit Aristotle's definition. Corresponding to the practical joke is Kant's account of the cause of laughter as "the sudden transformation of a strained expectation into nothing" (as when a playful adult offers an infant a toy and then suddenly yanks it back with mock laughter). The Kantian definition fits adult humor, where often the funny thing is that "a plausible movement of the mind comes to nothing".[9] Much comedy and perhaps all wit, Eastman tells us, have this Kantian character which in turn is a refinement of the infant's gleeful snatching at the withdrawn toy, in this case a withdrawn meaning. "A trick is played upon our minds – a playful disappointment of their momentary expectation, and a pleasure offered to their underlying trends".[10] Some of the most exquisite comedy is directed at patently absurd reasonings and ludicrously incongruous situations,[11] but only when we can take them playfully, only, that is, when distracting anxieties, fears, and such emotions as anger, sorrow, compassion, and indignation, have been 'paid off' and sent on a holiday. In short, an element of absurdity in an unhappy situation is itself a comic element in that situation no matter how unhappy the whole complex of which it is an element, but that is only to say that *if* we could isolate that element, abstract it from its whole complex and focus narrowly on it in a wholly playful spirit, we would find it amusing. (Why that should be the case even Eastman does not wholly explain.)

II. THE ABSURDITY OF HUMAN LIFE

Taylor, Camus, and Nagel, each in his own way and each making his own special qualifications, looks with favor on the judgment that there is absurdity in the human condition as such. It will be useful here to summarize the reasons given by these authors. We can begin with Taylor, who finds all human activity to be as pointless (in the long run) as that of Sisyphus. He uses the words 'meaningless' and 'pointless' instead of 'absurd', and as we have seen, means by them "endless repetitive activity that comes to nothing". The 'endlessness', no doubt, is inessential to the meaninglessness. If Sisyphus pushed his rock continuously for four score and ten years only, before being mercifully killed by the gods, Taylor could and would judge his finite career as a land-mover to be absurd, just as he judges the finite lives of men and mice to be absurd. Meaninglessness for Taylor is mitigated but not canceled by achievement, because achievements do not last. Some achievements, for example *Hamlet,* Beethoven's *Fifth Symphony*, and the Notre Dame cathedral, last longer than others, and might therefore qualify as less absurd than the transient and trivial triumphs in which most perhaps take what pride they can. But from any sufficiently broad point of view, long compared to the span of human lives or even the lives of nations and planets, but infinitely narrower than the perspective *sub specie aeternitatis*, the difference between the durability of Notre Dame and that of a pioneer's log cabin is utterly insignificant. All of our goals, Taylor says, are of "transitory significance", and "having gained one of them we immediately set forth for the next, as if that one had never been, with this next one being essentially more of the same".[12] Unlike Sisyphus, however, most of us beget children and pass on our values, our modest achievements, and fresh opportunities to them. That fact does not impress Taylor, who replies that "Each man's life thus resembles one of Sisyphus's climbs to the summit of his hill and each day of it one of his steps; the difference is that whereas Sisyphus himself returns to push the stone up again, we leave this to our children".[13] The enterprise is thus collective, but it still comes to nothing in the end.

What could human existence conceivably be like if it were to escape this absurdity? Taylor has us imagine that Sisyphus is permitted to push different stones to the top of his hill and combine them there into a beautiful and enduring temple. This would be different, Taylor says at first, for "activity . . . has a meaning if it has some significant culmination, some more or less lasting end that can be considered to have been the direction and purpose of the activity".[14] But soon he changes his mind. He does not wish to make meaningfulness a matter of "more or less", for then he would have to admit that some human activities and lives are to some degree, at least, meaningful, and comparisons of individual lives in terms of their relative meaningfulness would at least make sense. But that would be to vindicate rather than to destroy common sense on this question, and Taylor quickly withdraws his concession by requiring that the temple must endure – not simply be "more or less lasting" – "adding beauty to the world for the remainder of time".[15] When we look at a meaningless life like that of the legendary Sisyphus or that of a drug-addicted teenage suicide, and compare it with one of the relatively meaningful human lives suggested by common sense, say that of Jefferson or Shakespeare, the differences at first, are striking. But "if we look at them from a distance" (say from a point in time one hundred million years from now) they "are in outline the same and convey to the mind the same image" – pointless labor coming to nothing.[16] It is the temporal distance that makes the difference. The view from remote distances in time reveals things as they truly are, whereas the detailed close-up picture is distorted and illusory.

Taylor has one final supposition to make. Let us suppose that after a finite period of intense labor Sisyphus finishes a gloriously beautiful temple, and then is allowed by the gods to rest on his laurels and then spend the rest of eternal time in admiring contemplation of his significant achievement. Now at last we seem to have an unchallegeable conception of nonabsurd existence, but Taylor quickly dashes our hopes. Eternal rest, he rightly claims, would be "infinite boredom", and that too would become in due time a kind of pointlessness or absurdity. He does not consider still other possible suppositions: that Sisyphus be allowed to die after a brief rest period but before his proud satisfaction turns to boredom, while his temple is preserved forever by the gods, or that Sisyphus be permitted to live forever, alternating creative activity with replenishing periods of rest, while the gods guarantee the permanence of his achievements. I suspect that Taylor, like Bernard Williams,[17] would find even the latter arrangement no escape from infinite boredom, so that his final view, if he had finished his argument, would be that almost any conceivable form of human life would be absurd, either because it fails to produce lasting achievements or because it leads to infinite boredom.

According to Albert Camus, human beings necessarily crave a certain kind of cosmic order, significant culminations of their efforts, and a kind of transparent rational intelligibility in the world of experience. But the world has no such order; it works to destroy the point of whatever temporary achievements it permits; and it is in its central core alien, dense, and irrational. Hence the inevitable confrontation and the inevitable absurdity. What should we do about it? Suicide would itself be a pointless gesture. Self-deception is the more common way out. But embracing consoling myths is inconsistent with one's integrity. There is in fact no way of reconciling the cravings inherent in our nature with the uncompromising denials of the alien cosmos. The existentialist hero acknowledges his inherent absurdity without wincing; he cherishes his consciousness of it, keeping it forever alive, as the evidence of his integrity. He has no hope that things could be different, but lives to the hilt and dies well, like a blind person who cannot relinquish his desire to see though he knows the desire is hopeless.[18] In his defiance of what is necessary he achieves his integrity, and in his revolt, his happiness.

Thomas Nagel presents some undeniable examples of absurdity *in* life before raising the subject of the absurdity *of* life. Since the examples are indeed good ones and such that they are easy to respond to playfully, they are irresistibly comic: "Someone gives a complicated speech in support of a motion that has already been passed; a notorious criminal is made president of a major philanthropic foundation; you declare your love over the telephone to a recorded announcement; as you are knighted, your pants fall down."

What these and similar examples from ordinary life have in common, Nagel tells us, is "a conspicuous discrepancy between pretension or aspiration and reality".[19] Surely Nagel is right about that, and his examples are right on the mark, but one wonders how useful his instances of absurdity in life are as models for the alleged absurdity of all human lives. Applied to Sisyphus, at any rate, Nagel's "discrepancy between pretension or aspiration and reality" seems less relevant than Taylor's conception of ultimate futility and pointlessness.

Nagel's type of absurdity would add another element to the pointlessness of Sisyphus' labors. Imagine for example that the gods have sentenced Sisyphus only to climb a large mountain and plant a small flag on the top. As soon as he succeeds in doing that, his penalty has been paid once and for all. It takes Sisyphus years (or centuries) to climb the mountain but then he finds that he has forgotten the flag. He returns to the base, recovers the flag, and spends another millennium or so climbing a peak only to discover that he is on the wrong mountain. And so on *ad infinitum*. Sisyphus' labors would not be pointless in that case since they would have a sensible aim, but how genuinely absurd he would be! Or suppose that the gods in the original legend had not only assigned Sisyphus his endless self-defeating labor, but had also required him, before each trip to the summit, to write 'I am a bad boy' one hundred times on his rock. What an absurd comedown for the proud and once mighty Sisyphus! How discrepant with his pretensions!

Moreover, traditional Sisyphus, before we began tampering with the legend, was *not* absurd on Nagel's definition. He had no illusions or false pretensions, and his resigned 'aspiration' was perfectly realistic. If his plight, therefore, is to be taken as a model for the absurdity of the whole human enterprise, we shall have to expand Nagel's account of absurdity to include examples from within human life of the sort that Taylor emphasized, for instance writing novels and then burning them, ordering intricately ornate wedding cakes and then consuming them before they are even wrapped, or in general, giving with one hand and taking away with the other.[20] If in fact we all are like Sisyphus, however, and we pretend to be otherwise, then we are absurd in Nagel's sense too.

The absurdity in the human condition, according to Nagel, derives from the clash of perspectives from which we can view ourselves: that of purposeful actors living out our lives and that of disinterested spectators of the very lives we earnestly live. Only human beings are capable of viewing themselves from a detached and impersonal perspective and of making judgments from that viewpoint of their own insignificance, and when we do view ourselves in that detached way, then the ordinary way of regarding our lives, which we cannot help but adopt if we are to pursue our lives at all, seems absurd to us.

A mouse also regards his own life in the same serious everyday way that humans do, but since "he lacks the capacities for self-consciousness and self-transcendence that would enable him to see that he is only a mouse",[21] he is not absurd. Human beings can diminish (but probably not eliminate) the absurdity of their own lives by allowing their individual animal natures to drift and respond to impulse, in short by becoming as much like mice as possible, but this would involve "considerable dissociative cost".[22]

Nagel gets confused, or at least confusing, when he talks as if he is making judgments about the absurdity of others' lives when he is only explaining the way in which those lives might come to *seem* absurd, either to those persons themselves or to a sensitive observer. Thus when he talks about possible 'escapes' from absurdity and admits that a mindless life spend drifting with impulse is less absurd than more characteristically human lives despite its dissociative cost, he is using 'is absurd' to mean 'seems absurd', much as psychoanalysts often equate 'guilt' and 'feelings of guilt'. The life of a mouse *is* absurd when we look at it from an imaginatively extended perspective that the mouse itself cannot achieve. When Nagel denies the mouse's absurdity on the ground that it has no transcendental consciousness, he explains why the mouse's life cannot *seem* absurd to it. But the mouse's life can still seem absurd to us, and really be absurd nonetheless. Nagel, in short, at least in much of his discussion, makes the essential discrepancy in an absurd life to be a relation between two components of the being whose life it is – his natural and inevitable seriousness, and his awareness from a higher perspective of his own insignificance. But one could lack that kind of discrepancy, as mice do, and enjoy a more unified consciousness which in turn is discrepant with an *external* reality, the unaccommodating and alien universe. Nagel employs the latter conception of absurdity too when he speaks of the clash between subjective pretension and objective reality, and that is the notion that is used by Taylor and Camus when they make their judgments of real, not merely apparent, absurdity.

The distinction between really being and only seeming absurd quickly suggests another, that between absurdity as a property of one's situation, and absurdity as a flaw in one's outlook or self-assessment; put tersely, between *absurd predicaments* and *absurd persons*. It does not follow, of course, from the fact that a person is an absurd predicament that he is himself an absurd person, for he may have redeeming insights into, and attitudes toward his situation that put him beyond criticism or mockery. The human predicament that we all share is absurd according to Taylor because achievements do not last and there is thus a necessary and objective discrepancy between effort and outcome. It is absurd according to Camus because the universe is resistant to our inherent craving for order and intelligibility and there is thus an ineradicable disharmony between our needs and the world's indifference. The human predicament is absurd according to Nagel because of the irresolvable clash between the importance we attach to our lives and the essential dubitability of all schemes of justification for that importance. All three writers agree that the absurdity of our human predicament is not a matter of 'more or less' and not a matter that could be different from what it is. It is otherwise with the absurdity of persons. Some people are obviously more absurd than others in that there is a greater clash between their beliefs and their evidence, their ideals and their practice, their means and their ends, or their pretensions and their real characters and situations.

A person is also absurd – and this is the interesting point – when there is a radical discrepancy between his assessment of his situation and the actual nature of that situation. If one is really in an absurd predicament, if for example all of one's labors are bound to come to nothing in the end whatever one chooses to do about it, and one stubbornly denies that absurdity, adopting inappropriate attitudes and embracing vain hopes, then one becomes more than a little absurd oneself. Thus Sisyphus escapes personal absurdity by correctly appraising the absurdity of his predicament, realistically abandoning hope, and coolly proceeding with his labors in an existentialist spirit of "Let's get on with it",[23] thus maintaining a kind of dignity and self-respect. But Sisyphus would surely be absurd if, like Don Quixote, he talked himself into believing that his labors had an intrinsic worth and importance and were essential to the maintenance of the world order. Indeed, we could imagine a number of possible Sisyphuses varying in their degree of personal absurdity or unabsurdity as their beliefs, assessments, attitudes, and pretensions vary in their degrees of fittingness to their predicament. The situation of all these hypothetical Sisyphuses, however, is the same and as thoroughly absurd as a situation can be, for whatever any Sisyphus chooses to do about it, he must engage in endless repetitive cycles of pointless and unproductive labor.

How can a person be unabsurd if his life as a whole is unavoidably absurd? Some self-attitudes do not further anyone's escape from absurdity, and in the case of the person whose situation itself is absurd and whose projects and enterprises are pointless, they positively accentuate the personal absurdity of their possessor. Vanity, excessive pride or shame, pompous self-importance, even well-grounded self-esteem if taken too seriously, are absurd in a person whose situation guarantees the pointlessness and futility of his activities. Think of Shelley's Ozymandias, for example, who built a monument for posterity directing his descendants to "Look on my works, ye Mighty, and despair!" A tick of cosmic time later only "Two vast and trunkless legs of stone/Stand in the desert. . . . Near them, on the sand,/Half sunk, a shattered visage lies, . . ." How absurd was old Ozymandias, self-declared "king of kings"![24] Equally absurd would be the towering self-regard of an eminent physicist for having won (and deserved) the Nobel Prize. Think of his proud medal found on some desert of the next millennium by beings whose schoolchildren have a far more advanced understanding of physics than he did. A little bit of genuine humility, perhaps, is a virtue in anyone in any situation, but for a person in an absurd situation it is essential if the absurdity of his predicament is not to rub off on his character.

III. THE CONCEPT OF SELF-FULFILLMENT

There are various technical concepts of self-fulfillment associated with the writings of such philosophers as Plato, Aristotle, Rousseau, and Hegel, whose histories go back to the earliest beginnings of Western philosophy and are equally venerable in Eastern thought. There also seem to be one or more notions of self-fulfillment, perhaps less clearly conceived and articulable, that are part of ordinary thought, as for example when people say that one kind of life, or one kind of marriage, hobby, or career, is preferred because it is more fulfilling.

In applying the ordinary concept of self-fulfillment, people seem, on different occasions, to use as many as four different models for their understanding. On the first model, fulfillment is simply the answering to *any* anticipatory condition, whether one's own or another's – promises, hopes, expectations, desires, requirements, or whatever. The second is 'filling up, being made full'. The third is the opposite of the second, namely emptying, unwinding, discharging, untying – draining one's cup of life to the dregs. Each of these familiar models comes with its own metaphors to guide (or obscure) the understanding. It is the fourth model, however, that of 'doing what comes naturally', that purports to be more 'philosophical', and is the more important one for our present purposes. This model restricts itself to the basic dispositions of one's 'nature', and where these differ or conflict, to the 'higher' or 'better' ones. Moreover, fulfillment on this model is not merely a discharging, but also a maturing and perfecting of our basic dispositions. Finally, fulfillment so interpreted is often said to be a 'realizing of one's potential', where the word 'potential' refers not only to one's basic natural proclivities to engage in activities of certain kinds, but also to one's natural capacities to acquire skills and talents, to exercise those abilities effectively, and thus to produce achievements. Insofar as one fails to 'realize one's potential', one's life is thought, on this as well as the third model, to be 'wasted'.

This understanding of self-fulfillment is much too general to be useful, and the main challenge to the philosophers, from Aristotle on, who have tried to incorporate it, has been to give it specificity. Almost anything one does can be said to fulfill a prior disposition to act in precisely that way in circumstances of that kind, or to implant or strengthen the habit of acting that way in the future. Thus almost any action can be said to discharge a natural tendency, to be a doing of what it is in one's nature to do. Philosophers who have fashioned a technical concept of self-fulfillment from the vaguer everyday notion have for the most part assigned it a crucial role in the definition of 'the good for man'. For that reason, most of them have begun the task of specification by ruling out as self-fulfilling, actions that violate objective standards of morality or which are radically defective in other ways. If a man has the bad habit, acquired and reinforced over a lifetime, of stealing purses, then a given act of stealing a purse, even though it fulfills one of the basic dispositions of his (evil) character, cannot be allowed to count as self-fulfilling. The same kind of fiat has excluded evil actions that discharge native propensities, for example the angry tirades or physical assaults of a person who is irascible, hotheaded, or aggressive 'by nature'. Such arbitrary exclusions do not shock common sense, but there does seem to be at least as much warrant in ordinary conceptions for saying that it may be a bad thing that certain kinds of self be fulfilled, but that the discharging of basic 'evil' dispositions remains fulfillment, and properly so called, anyway.

Some philosophers in the grand tradition have also excluded from their conception of self-fulfillment, activities that fulfill dispositions peculiar to individual persons, so as to give special importance to activities that fulfill those dispositions that define our common human nature. The phrase 'a person's nature' is of course ambiguous. It may refer to the nature he shares with all and only human beings, his 'generic nature' as it were, the nature that makes him classifiable as the kind of being he is, or it may refer to the nature that belongs uniquely to him, his 'individual nature', the character that distinguishes him from all other individuals of his kind. My generic nature includes my disposition to walk upright and to speak a language,[25] among other things. It is part of *my* individual nature, on the other hand, to be interested in philosophy, to be punctual at meetings, to be slow at mathematics, and to be irritable when very tired or hungry. Some of the traits that characterize me but not everybody else are not thought to be part of my individual *nature* because they are weak and tentative habits rather than governing propensities, or because they are trivial (like my habit of scratching my head when deliberating). My individual nature is partly acquired; my generic nature is derived entirely from heredity. I come into existence with it already 'loaded and cocked'.

Those philosophers of fulfillment who attach special significance to our generic natures tend to draw heavily on biological as well as mechanical metaphors. In a fulfilled life our preprogrammed pontentialities 'unfold' like the petals of a rose, each in its time, until the plant is fully flowered and 'flourishing'. Then there follows an equally natural, gradual withering and expiring, and the life of one plant, at least, has been fulfilled. Another plant, much like the first, is caught in a frost and nipped in the bud, never to achieve its 'own good' as determined by its natural latencies – the very paradigm of a tragic waste. John Stuart Mill referred to qualities which are "the distinctive endowment" not of the individual in question but of a human being as such: "the human faculties of perception, judgment, discriminative feeling, mental activity, even moral preference",[26] these understood as standing to human nature in the same relation as that in which unfolding and flourishing stand to the nature of a rose.

I believe it is a mistake, however (and not one committed in common thought) to exclude individual natures from one's conception of self-fulfillment. If we are told by philosophical sages to act always so as to unfold our generic human natures, we have not been given very clear directions at all. Any number of alternative lives might equally well fulfill one's generic nature, yet some might seem much more 'fulfilling', in a perfectly ordinary and intelligible sense, than others. William James made the point well:

> I am often confronted by the necessity of standing by one of my empirical selves and relinquishing the rest. Not that I would not, if I could, be both handsome and fat and well dressed, and a great athlete, and make a million a year, be a wit, a *bon vivant*, and a lady-killer, as well as philosopher, a philanthropist, statesman, warrior, and African explorer, as well as a "tone-poet" and saint. But the thing is simply impossible. The millionaire's work would run counter to the saint's; the *bon vivant* and the philanthropist would trip each other up; the philosopher and the lady-killer could not well keep house in the same tenement of clay. Such different characters may conceivably at the outset of life be alike possible to a man. But to make any one them actual, the rest must more or less be suppressed.[29]

All James's possible careers might equally well fulfill his human nature, just as all the variously colored unfoldings of roses might equally well fulfill a rose's generic nature, but a rose cannot pick its own individual character, whereas a man has some choice. Since some of James's lives (presumably the philosophical one, to begin with) would be more fulfilling than others, it must be his individual nature *qua* William James that makes that so. The point would be even clearer if James had listed among the possibilities, 'anchorite monk', 'operatic *basso profundo'.* 'brain surgeon', and 'drill sergeant'. Some of these careers obviously accord more closely than others with *anyone's* native aptitudes, inherited temperament, and natural inclinations. How does one choose among them if one is seeking fulfillment? By 'knowing oneself', of course, but not simply by knowing well the defining traits of *any* human being. To be sure, making the choice itself is a characteristically human act and calls into play all of the generic human traits on Mill's list – perception, discrimination, insight, and the like – but to exercise those traits effectively and well, and thus unfold one's generic human nature, one must first know one's individual character as so far formed, and make the decision that best *fits* it. Mill's final and favorite metaphor, indeed, is that of a life fitting an individual nature in the way a shoe fits a foot: "A man cannot get a coat or a pair of boots to fit him unless they are either made to his measure or he has a whole warehouseful to choose from; and is it easier to fit him with a life than with a coat . . . [?] "[28]

Some of a person's individual nature is native, for example much of what we call aptitudes, temperamental dispositions, and physical strength. A fulfilling life therefore is one that 'fits' these native endowments. But we make our own natures as we grow older, building on the native base. We begin, partly because of our inherited proclivities and talents, to develop tastes, habits, interests, and values. We cultivate the skills that grow naturally out of our aptitudes, and as we get better at them we enjoy them more and exercise them further so that they get better still, while we are inclined to neglect the tasks for which our skills are inadequate, and those abilities wither and decay on the vine.[29] The careers we then select as workers, players, and lovers, should be those that fit our well-formed individual natures, at least insofar as each stage in the emergence of the self grew naturally out of its predecessor in the direction of our native bent.

Emphasis should be given to the further point that fulfillment of one's generic and individual natures are interconnected and interdependent. The passages in *On Liberty* quoted above in which Mill urged fulfillment of the "distinctive endowment" of generic humanity occur, ironically, in a chapter entitled 'Individuality as One of the Elements of Well-Being', and nowhere in that chapter can Mill discuss individuality for long without bringing in human nature and vice versa. His view clearly was that it is essential to the generic nature of human beings that each think and decide for himself rather than blindly follow all the rest, so that in cultivating the capacities that human beings share in common, each individual will at the same time be promoting his own distinctive individuality. If I pick a career that fits my individual nature instead of blindly drifting with custom or passively acceding to the choices of another, then I have exercised my generic nature as a thinker and chooser, at the same time that I have promoted the fulfillment of my individual nature as a person with a unique profile of interests and aptitudes.

Useful as it may be for some purposes the distinction between generic and individual natures is vague and ragged about the edges, a point we can appreciate by returning to the plight of Taylor's Sisyphus. Depending on the extent to which the gods had to tamper with him, he has either had a new individual nature grafted on to his basic human nature, or else a new (hence nonhuman) generic nature installed in him. If we say the former, then we must think of his infinite rock-pushing proclivities as merely personal eccentricites, only contingently unshared by other persons who share his human nature. If we say the latter, then the individual nature of Sisyphus and his generic nature coincide, since he is now one of a kind, the sole member of his new species. A rock-pushing instinct that is so specific would be such a departure from what we normally think of as human nature, so totally unshared by any other humans that perhaps there would be a point in saying that Sisyphus has a new generic nature, humanoid but not human, and that he is now the only member of the biological species *Homo sapiens geopetris* – sapient rock pusher. Still more plausible, perhaps, we might think of the new Sisyphus as a borderline case for our old classifications. Unless we hold to the discredited doctrine of fixed species, we can simply declare that there is no uniquely correct answer to the question of whether Sisyphus' generic nature has been changed, and that considerations of convenience and tidiness are as relevant to its resolution as are any questions of fact.

Moreover, when we consider thoughtfully the whole range of hypothetical Sisyphuses from which we might draw in order to flesh out Taylor's example, we are struck with how very vague the notion of a 'nature' is, whether generic or individual. What is in Sisyphus' nature (or the nature of anyone else) is very much like what is in his (cluttered) closet or in his grab bag, including everything from aptitudes and interests to addictive compulsions. Think of all the variations on Taylor's theme: the gods might have implanted in Sisyphus an *appetite* for stone-pushing which makes regular and frequent demands on him, like hunger or the 'sex urge' in others, and corresponding in its cycles to the time it takes to push a rock of standard size up Sisyphus' assigned mountain and then return again to the bottom of the hill. 'Ye gods!', he might exclaim after each round of labors, 'how I hunger for a nice big rock to push', and the accommodating deities always have one ready for him, like the next ball up in a pinball machine. Or the gods might have designed for him a peculiar *talent* for rock-pushing much like others' talents for piano-playing, tennis, or chess. The new Sisyphus starts all over as a perpetual youth, and from the start he is a veritable prodigy at rock-pushing. He comes to enjoy exercising his skills, and makes ever-new challenges for himself. He pushes the rock right-handed, then left-handed, then no-handed, then blindfolded, then does two at a time, then juggles three in the air all the way to the summit, eager to return for another rock so that he can break his record, or equal it next time while dancing a Grecian jig. Or the gods do their job by implanting an *instinct* for rock-pushing so that Sisyphus goes about his chores without giving them so much as a thought (except in rare reflective moments and then only to shrug his shoulders and get on with it). His work is as natural and unremarkable to him as having a language or standing upright is to us, or building a dam to a beaver, or peeling a banana to a chimpanzee. Or (perhaps more plausibly) the gods implant a *drive* or more general proclivity of which stone-pushing is only one of numerous possible fulfillments. If there were only opportunity to do so, Sisyphus would find it equally in his nature to push wooden logs, or plastic bags, or iron bars, or to pull, lift, carry and throw objects or push them while swimming against a current, or to pile, hook, or nail them on to one another as in construction work, and the like. But pushing rocks up a mountain will do as well as any of the other activities as fulfillments of his drive to move and manipulate physical objects and he can be grateful to the gods for that. Or, the gods can use Taylor's own suggested method, and give Sisyphus (say) a shot in the arm after each trip so that he will feel a "*compulsive impulse*" to push the rock up once more in order to get relief in the form of another addicting shot. This technique would keep the gods busier than the others, but they could let some internal gland, timed to secrete the essential substance into Sisyphus' bloodstream at appropriate intervals, do the work for them. There is something especially ingenious in this last scheme, for the 'shot' given at the base of the hill creates the impulse to push the rock up the hill and also the addictive need, when its first effect wears off, to be renewed by another shot, and so on, *ad infinitum.*

If the gods' gift to Sisyphus is merely an *appetite* to push rocks he may yet fail to find self-fulfillment on balance in an indefinitely extended lifetime of rock-pushing, just as one of us might fail to be fulfilled in a life that gives us all the food we need, but nothing else. Sisyphus will have the periodic satisfactions of regular appetite-satiation, and that is certainly some benefit to him, but the deepest yearnings of his nature will nevertheless be forever denied. Much the same can be said of his condition if the gods simply addict him chemically to a substance that creates a rock-pushing itch, or if they implant in him an extrahuman instinct to push rocks that fails to dovetail or integrate with the human instincts he must continue to maintain if he is to preserve his identity with his earlier self. The model that makes talk of Sisyphean self-fulfillment most plausible is probably that in which the gods impart to him *talents* for rock-moving that he can forever after exercise and glory in. So endowed, he can find self-fulfillment through his developed virtuosity, in the same way others find fulfillment in lives of skilled celloplaying or cabinetmaking.

No conception of self-fulfillment will make much sense unless it allows that fulfillment is a matter of degree. We begin life with a large number of potential careers some of which fit our native bent more closely than others but any of which, if pursued through a lifetime, would lead to substantial fulfillment, so that the pursuit of no *one* of them is indispensable to a fulfilled life. Imagine a warm and loving woman who is superbly equipped by her nature to be a parent, and has thought of herself throughout her girlhood as a potential mother. She marries and then discovers that she is barren. Had she not been infertile she would have achieved fulfillment in a long lifetime of nearly full-time motherhood. Is it now impossible for her to be fulfilled in a life without children? Clearly not, for the very traits that make her 'superbly equipped' to be a mother will make her more than a little qualified for dozens of other roles, and a fulfilled life could stem from any one of these, from social work to schoolteaching,[30] or even from a career based on independent specific aptitudes like poetry or basket weaving. She may be disappointed that her chief ambition is squelched and her regrets may last a lifetime, but disappointment and fulfillment can coexist with little friction, as they do to some degree in most human lives. Thus, we each have within us a number of distinct individual possibilities, several (at least) of which would be sufficient for (a degree of) fulfillment, but no one of which is necessary. But the most fulfilling ones are those that best *fit* one's latent talents, interests, and initial bent and with one's evolving self-ideal (as opposed simply to one's conscious desires or formulated ambitions).

Some fulfilled human lives are relatively monochromatic, having a single dominant theme; others are diversely colored, having a harmonious orchestration of themes with equal voices. All of them approach fulfillment insofar as they fill their natural allotment of years with vigorous activity. They need not be 'successful', or 'triumphant', or even contented on balance in order to be fulfilled, provided they are long lifetimes full of struggles and strivings, achievements and noble failures, contentments and frustrations, friendships and enmities, exertions and relaxations, seriousness and playfulness through all the programmed stages of growth and decay. Most important of all, a fulfilled human life will be a life of planning, designing, making order out of confusion and system out of randomness, a life of building, repairing, rebuilding, creating, pursuing goals, and solving problems. It is in the generic nature of the human animal to address the future, change its course, make the best of his situation. If one's house falls down, if one's cities are in rubble, if disaster comes and goes, the human inclination is to start all over again, rebuilding from scratch. There is no 'fulfillment' in resignation and despair.

Sisyphus does seem very human after all, then, when he reshoulders his burden and starts back up his hill. But insofar as his situation is rigidly fixed by the gods, allowing him no discretion to select means, design strategies, and solve problems on his own, his life does not fulfill the governing human propensities. If he can fulfill his nature without these discretionary activities, then he has really assumed the nature of a different species. In all the variations on the Sisyphus myth that we have spun thus far, the gods have assigned a very specific job to Sisyphus that requires no particular judgment or ingenuity on his part to be performed well. They have imposed a duty on him rather than assigning a responsibility.[31] He has a rote job to perform over and over, a mulish task for a mulish fellow, and his is not to reason why or how, but only to get on with it. Suppose, however, that the gods assign to Sisyphus an endless series of rather complex engineering problems and leave it up to him to work out how to solve them. Somehow rocks must be moved to mountain tops and there can be no excuses for failure. 'Get it up there somehow', they say; 'the methods are up to you. Feel free to experiment and invent. Keep a record of your intermediate successes and failures and be prepared to give us an accounting of the costs. You may hire your own assistants and within certain well-defined limits you have authority to give them commands, so long as you are prepared to answer for the consequences of their work. Now good luck to you." If Sisyphus' subsequent labors are fulfilling, they will be so in a characteristically human way. His individual nature will be fulfilled by a life (endless and pointless though it may be) that fits his native bent and employs his inherited talents and dispositions to the fullest, as well as fitting his more specific individual tendencies, for example a special fascination (perhaps also a gift of the gods) with rocks.

V. WHY DOES SELF-FULFILLMENT MATTER?

Why should it 'matter' that a person is unfulfilled if, despite his stunted and dwarfed self, the product perhaps of alienating work and other 'unfitting' circumstances, he finds a steady diet of satisfactions in delusory occupations, escapist literature, drugs, drink, and television? Why should it 'matter', to turn the question around, that a person finds fulfillment when his life looks as absurd from a longer perspective as the life of a shellfish appears to us?

Think first of what a substantially unfulfilled life involves. A person comes into existence with a set of governing dispositions that set him off with others as a being of a certain kind. For twenty years he grows and matures, enlarging and perfecting his inherited propensities so that he becomes utterly unique, with a profile of talents and individual traits which, as a group, distinguish him from every other being who has ever existed, and constitute his individual nature. Perhaps he is capable of seeing, from time to time, that this 'nature' of his is more than a little absurd. What he does best and most, let us imagine, is play chess and ping-pong and socialize with others who share those interests. He takes those pursuits more seriously than anything else in his life. But he knows that they are, after all, only games, of no cosmic significance whatever, and certainly of no interest to the indifferent Universe, to Posterity, to History, or to any of the other abstract tribunals by which humans in their more magniloquent moods are wont to measure significance. And yet, absurd as it is, it is *his* nature, and the only one he has, so somehow he must make the best of it and seek his own good in pursuit of *its* dominant talents. Whose nature could he try to fulfill, after all, but his own? Where else can his *own* good conceivably be found? It was not up to him to choose his own nature, for that would presuppose that the choosing self already had a nature of its own determining its choice. But given the nature with which he finds himself indissolubly identified for better or worse, he must follow the path discovered in it and identify his good with the goals toward which his nature is already inclined.

Now suppose that he makes a mess of it through imprudence, frivolity, or recklessness; or imagine that the world withdraws its opportunities; or that lightning strikes and leaves him critically incapacitated for the realization of his potential. That leaves him still the pleasure of his diminished consciousness, his soma pills and television programs, his comic books and crossword puzzles, but his deepest nature will forever remain unfulfilled. Now we think of that nature, with all of its elaborate neurochemical equipment underlying its distinctive drives and talents and forming its uniquely complex character, as largely *unused, wasted, all for naught*. All wound up, it can never discharge or wind down again. In contrast, the life of fulfillment strikes us as one that comes into being prone and equipped to do its thing, and then uses itself up doing that thing, without waste, blockage, or friction.

When any nature is left unfulfilled it is likely to strike us as a bad thing, an objectively regrettable fact. Perhaps we would withdraw or modify that judgment when we come to appreciate how absurd that nature's preoccupations really were. But from the point of view of the self whose nature it is, nonfulfillment is more than a bad or regrettable thing to be graded down in some negative but modifiable 'value judgment'. It may or may not be all those things in some final balancing-up, all things considered. But from the point of view of the individual involved, nonfulfillment marks the collapse of his whole universe, the denial once and for all of his *own good*. There is a world of difference in the use of the word 'good' as a predicate of evaluation, and its use in the venerable phrase of the philosophers – 'one's own good'. My good is something peculiarly mine, as determined by my nature alone, and particularly by its most powerful trends and currents. Anything else that is *good for me* (or in my interest) is good because it contributes to my good, the fulfillment of my strongest stable tendencies. One can judge or evaluate that good from some other standpoint, employing some other standard, and the resultant judgment may use the words 'good', 'bad' or 'indifferent'. It may not be a good thing that my good be achieved or that it be achieved in a given way, or at a given cost. But it is logically irrelevant to the question of *what my good is* whether my good is itself 'good' when judged from an external position. My nonfulfillment may not be a 'bad thing on balance' in another's judgment or even in my own. My nonfulfillment may not be 'objectively regrettable' or tragic. But my nonfulfillment cannot be *my good* even if it is from all other measuring points, a good thing.

It is perhaps not quite self-evident that my good consists in fulfillment. A hedonist might hold out for the position that my good consists in a balance of pleasant over unpleasant experiences while denying that the basic disposition of my nature is to seek pleasure, thus denying that pleasant experiences as such are fulfilling. I cannot refute such a heroic (and lonely) philosopher. But I would like to urge against the philosopher who is overly impressed with the fact of human absurdity, that *if* my good is fulfillment, it must be fulfillment of *my* nature and not of something else. That my nature is eccentric, absurd, laughable, trivial, cosmically insignificant, is neither here nor there. Such as it is, it is my nature for better or worse. The self whose good is at issue is the self I am and not some other self that I might have been. If I had had any choice in the matter I might have preferred to come into existence with the nature (that is the potential) of William James, John F. Kennedy, or O. J. Simpson, but I cannot spend all my days lamenting that the only nature whose fulfillment constitutes my good is my own!

The prerequisite to self-fulfillment is a certain amount of clear-eyed, non-deluded *self-love*. A moment ago I spoke of one's own nature 'such as it is', 'for better or worse'. These phrases recall the wedding ceremony and its conception of marital love as loyalty and devotion without condition or reservation. *Totally* unconditional devotion may be too much to ask from any lover, but within wide limits, various kinds of human love of others have a largely unconditional character. Gregory Vlastos describes parental love, for example, in a way that makes it quite familiar: "Constancy of affection in the face of variations of merit is one of the surest tests of whether a parent does love a child".[32] Judgments of merit have nothing to do with love so construed. A child's failures, even moral failures, may disappoint his parents' hopes without weakening their loyalty or affection in the slightest. A parent may admire one child more than another, or like (in the sense of 'enjoy') one more than another, as well as judge one higher than the other, but it is a necessary condition of parental love that, short of limiting extremes, it not fluctuate with these responses to merit.

The love which any stable person has toward himself will be similarly constant and independent of perceived merits and demerits. I may (realistically) assign myself very low grades for physique, intellect, talent, even character – indeed I may ascribe deficiencies even to my individual nature itself – while still remaining steadfastly loyal and affectionate to myself. Aristotle was right on target when he said that a wise man ought to have exactly that degree of self-esteem that is dictated by the facts, neither more nor less. But self-esteem is not self-love. I have self-love for myself when I accept my nature as given, without apology or regret, even as I work, within the limits it imposes, for self-improvement. We have been through a lot together, my self and I, sharing everything alike, and as long as I have supported him, he has never let me down. I have scolded him, but never cursed his nature. He is flawed all right, and deeply so, but when the warts show, I smile, fondly and indulgently. His blunders are just what one would expect from anyone with his nature. One cannot come to hate a being with whom one has been so very intimate. Indeed, I would not know how to begin to cope with another self after all my years of dependency, 'for better or worse', on this one. In this way self-identity can be conceived as a kind of arranged marriage (I did not select the self that was to be me) that in a stable person ripens into true love, but in an unstable one sours into rancor and self-destruction. And the truest expression of one's self-love is devotion toward one's own good, which is the fulfillment of one's own (who else's?) nature absurd as that may be.

VI. THE CRITIQUE OF COSMIC ATTITUDES

Some lives are manifestly and incontrovertibly absurd. Lives spent moving metaphoric rocks back and forth to no further end and lives spent tangling with metaphoric windmills are cases in point. Other lives are full of achievement and design. In these lives, intermediate goals lend meaning to the pursuits that are instrumental to their achievement, and they in turn are given a point by the more ultimate goals they subserve. No goal is *the* ultimate one, however, for the most general ends are themselves means to a great variety of other ends, all tied together in an intricate and harmonious web of purpose. There may be no purpose to the whole web except its service to its own component parts, but each constituent has a place and a vindicating significance to the person whose life it is. Such a life is, relatively speaking, not absurd. There is no doubt an important practical point in distinguishing human lives in terms of their degree of absurdity, even in highlighting and emphasizing the distinctions. (Marx's doctrine of alienation is an example of the social utility of making such distinctions.) As we have seen, however, philosophers have found reason to claim that there is a kind of cosmic absurdity inherent in the human condition as such. As we stand back and look at ourselves from an extended temporal position, the distinction between absurd and nonabsurd lives begins to fade into insignificance, and finally vanishes altogether.

We also make useful distinctions between relatively fulfilled and unfulfilled selves, or fulfilling and unfulfilling lives. However we interpret 'fulfillment' – as the development of one's chief aptitudes into genuine talents in a life that gives them scope, or an unfolding of all basic tendencies and inclinations, or an active realization of the universally human propensities to plan, design, make order – there are wide differences among persons in the degree of fulfillment they achieve. Some lives are wasted; some are partially wasted and partially fulfilled; others are nearly totally fulfilled. Unlike the contrast between absurd and nonabsurd lives, these distinctions seem to be time-resistant. If Hubert Humphrey's life was fulfilling to him, that is a fact like any other, and it never ceases to be true that it was a fact. From any temporal distance from which it can be observed at all it will continue to appear to be a fact (though a diminishingly interesting or important one).

Consider a human life that is near-totally fulfilled, yet from a quite accessible imaginary vantage point is apparently absurd. Insofar as the person in question is fulfilled, he ought to 'feel good' about his life, and rejoice that he has achieved his good. Suppose that he realizes then how futile it all was, 'coming to nothing in the end'. What would be the appropriate attitude in that event to hold toward his life? Unchanged pride and satisfaction? Bitterness and despair? Haughty existential scorn? We can call such responsive attitudes taken toward one's whole life and by implication toward the whole human condition, 'cosmic attitudes'. One of the traditional tasks of philosophy (and what 'philosophy' is entirely about in the minds of innocent persons unacquainted with the academic discipline of that name) is to perform a kind of literary criticism of cosmic attitudes. It used to be the custom for philosophers not only to describe the universe in its more general aspects but to recommend cosmic attitudes toward the world as so described.

I welcome the suggestion of Thomas Nagel that the appropriate responsive attitude toward human lives that are both absurd and fulfilled is *irony,*[33] and I shall conclude by elaborating that suggestion somewhat beyond the bare recommendation that Nagel offers.

None of the familiar senses of irony in language or in objective occurrences seem to make any sense out of the advice that we respond to absurdity with irony. What Nagel had in mind clearly was another sense in which irony is a kind of outlook on events, namely, "an attitude of detached awareness of incongruity".[34] This is a state of mind halfway between seriousness and playfulness. It may even seem to the person involved that he is both very serious and playful at the same time. The tension between these opposed elements pulling in their opposite ways creates at least temporarily a kind of mental equilibrium not unlike that of the boy in Lincoln's story who was 'too scared to laugh and too big to cry', except that the boy squirms with discomfort whereas irony is on balance an *appreciative* attitude. One appreciates the perceived incongruity much as one does in humor, where the sudden unexpected perception of incongruity produces laughter. Here the appreciation is more deliberate and intellectual. The situation is too unpleasant in some way – sad, threatening, disappointing – to permit the relaxed playfulness of spirit prerequisite to the comic response. There *is* a kind of bittersweet pleasure in it, but not the pleasure of amusement. The situation is surely not seen as funny, though perhaps it would be if only one could achieve a still more detached outlook on it. One contemplates a situation with irony when one looks the facts in the eye and responds in an appreciative way to their incongruous aspects as such. Irony is quite different from despair-cum-tears, scornful defiance-cum-anger, and amusement-cum-laughter. It is pleasant enough to be expressed characteristically in a smile, but a somewhat tired smile, with a touch both of gentleness and mischievousness in it, as befitting the expression of a tempered pleasure.

In one of the most moving scenes of the twenty-seven-part B.B.C. documentary film on the First World War, a group of British reinforcements is shown marching toward the front. We know that they are cannon fodder marching to their own slaughter, and they know it too. They are foot-sore and bone-weary, and splattered with mud, and a steady rain is falling. The song they sing as they march is not a rousing anthem like *La Marseillaise* or *Rule, Britannia!*, not a cocky fight song like *Over There*, not a jolly drinking song like *Waltzing Matilda*, not a sentimental ballad, hymn of lamentation, or a mournful dirge. Instead they sing to the stirring tune of *Auld Lang Syne* the famous nonsense verse they created for the occasion:

We're here because we're here
Because we're here because we're here;
We're here because we're here
Because we're here because we're here . . .

The observer of the film feels a sudden pang and finds himself near tears, but quickly he perceives the absurdity in the lyrics and responds appreciatively to it, just as the troops, by selecting those words, are responding to the perceived absurdity in their situation. The sensitive observer sees how fitting the ironic response of the soldiers is, (and how dreadfully false any of its standard alternatives would have been) and himself takes a quiet sad pleasure in it. The soldiers were in an inescapably absurd predicament, without hope, and only by their unflinching acceptance of the absurdity of their situation are they saved from absurdity themselves. For us, the unseen audience, there is an inspiration in their example that makes the scene noble.

I do not mean to suggest for a moment that the march of the doomed soldiers is an apt metaphor for the whole of human life. The soldiers' brief lives were tragically wasted. If they had been specially-bred military animals they might have found both a personal and biological fulfillment in their peculiar demises, but they were ordinary humans whose bizarre and untimely deaths climaxed their undeveloped and unfinished lives. In contrast, many individuals do achieve fulfillment in long, active, creative lives. These lives are more than just 'worthwhile'; they represent to those who lead them the achievement of the only condition that can plausibly be deemed 'their good'. So philosophical 'Pessimism', the view of Schopenhauer and others that *no* life can *possibly* be worth living given the absurdity of the human condition, must be rejected. Its logical contrary, that cosmic Optimism that holds that all human lives necessarily are, or always can be, good and worthwhile, must also be rejected, in favor of the commonsense view that fulfillment requires luck, and luck is not always good in a world that contains violent passions, accidents, disease, and war.

In this essay, however, I have tried not only to sketch a conception of the good life and the bad but also to recommend an appropriate attitude toward the human condition generally. Imagine a person who both through his own virtues and good luck has led a maximally fulfilling life into his final declining years. He has realized his highest individual potential in a career that perfectly fit his inherited temperamental proclivities. His talents and virtues have unfolded steadily in a life that gave them limitless opportunity for exercise, and he has similarly perfected his generic human powers of discrimination, sympathy, and judgment in a life full of intermeshing purposes and goals. All of this is a source of rich satisfaction to him, until in the philosophical autumn of his days, he chances upon the legend of Sisyphus, the commentary of Camus, and the essays of Taylor and Nagel. In a flash he sees the vanity of all his pursuits, the total permeability of his achievements by time, the lack of any long-term rationale for his purposes, in a word the absurdity of his (otherwise good) life. At first he will feel a keen twinge. But unless he be misled by the sophistries of the philosophical Pessimists who confuse the empty ideal of long-term coherence with the Good for Man, he will soon recover. And then will come a dawning bittersweet appreciation of the cosmic incongruities first called to his attention by the philosophers. The thought that there should be a kind of joke at the heart of human existence begins to please (if not quite tickle) him. Now he can die not with a whine or a snarl, but with an ironic smile.

University of Arizona

NOTES

1 Richard Taylor, *Good and Evil*, Macmillan, London, 1970, p. 259.

2 *Ibid.*

3 The various essays in which Camus gives his most thorough account of absurdity have been translated into English by Justin O'Brien and published in one volume under the title *The Myth of Sisyphus and Other Essays*, Random House, New York, 1955.

4 Taylor, *op. cit.*, p. 258.

5 Albert Camus, *op. cit.*, pp. 22–23.

6 D. H. Parker, *The Principles of Aesthetics*, 2nd ed., Crofts, New York, 1947, pp. 94–102.

7 Thomas Nagel, 'The Absurd', *Journal of Philosophy* 68 (1971), 716–727.

8 The essence of the comic, according to Arthur Schopenhauer is "the sudden perception of . . . incongruity . . .", *The World as Will and Idea*, trans. by R. B. Haldane and

John Kemp, Trübner, London, 1886, Vol. 1, p. 76.

[9] Max Eastman, *The Enjoyment of Laughter*, Simon and Schuster, New York, 1936, p. 38.

[10] *Ibid*., p. 39.

[11] Eastman rightly insists on the separation of two distinct species of funny things, those that are funny to look at and those which take one by surprise, the former fitting the Aristotelian and the latter the Kantian model. "There are accordingly, two kinds of humor: taking an unpleasant presentation playfully and taking a frustrated thought or action playfully." Eastman points out that some writers use the term 'incongruous' for both comic types, and explains why: "Both practical and perceptual humor can be described loosely as incongruous, the one disrupting a temporal, the other a spatial pattern or congruity" (*op. cit*., p. 51).

[12] Taylor, *op. cit*., p. 262.

[13] *Ibid*., p. 263.

[14] *Ibid*., p. 260.

[15] *Ibid*., p. 263.

[16] *Ibid*., p. 264.

[17] Bernard Williams, "The Makropoulos Case: Reflections on the Tedium of Immortality", in *Problems of the Self*, Cambridge University Press, Cambridge, 1973, pp. 82–100. Jonathan Glover replies to Williams: "But I am not convinced that someone with a fairly constant character *need* eventually become intolerably bored, so long as [he] can watch the world continue to unfold and go on asking new questions and thinking, and so long as there are other people to share their feelings and thoughts with. Given the company of the right people, I would be glad of the chance to sample a few million years and see how it went" (*Causing Death and Saving Lives*, Penguin, Harmondsworth, 1977, p. 57).

[18] Camus, *op. cit*., p. 22.

[19] Nagel, *op. cit*., p. 718.

[20] Or first taking away and then giving, as in the unfunny example of absurdity from the Civil War: "Lincoln and Brooks lingered at the cot of a wounded soldier who held with a weak white hand a tract given him by a well-dressed lady performing good works that morning. The soldier read the title of the tract and began laughing. Lincoln noticed that the lady of good works was still nearby, and told the soldier that undoubtedly the lady meant well. 'It is hardly fair for you to laugh at her gift.' The soldier gave Lincoln something to remember. 'Mr. President, how can I help laughing a little? She has given me a tract on the "Sin of Dancing," and both my legs are shot off'." (Carl Sandburg, *Abraham Lincoln: The War Years*, Harcourt, Brace, New York, 1926, Vol. 2, p. 293).

[21] Nagel, *op. cit*., p. 718.

[22] *Ibid*., p. 726.

[23] The final line of Jean-Paul Sartre's play, *No Exit* (*Huis clos*), translated by Stuart Gilbert, A. A. Knopf, New York, 1946.

[24] Percy Bysshe Shelley, *Ozymandias* 2–4, 11, 12.

[25] There have been feral children who have permanently *lost* their ability to learn a language and children born without legs who never acquire the ability to walk. But insofar as these persons are human beings, they are born with the innate *capacity* to acquire the dispositions and skills involved in walking and talking even though circumstances prevent that capacity from being realized. The capacities in question are often

conditional ones: all human children have the capacity to learn a language, which is activated between the ages of two and twelve only, and only if they are made part of a language-speaking community during those years. That *conditional capacity* to acquire the dispositions and skills involved in language use *is* common to all human beings.

26 J. S. Mill, *On Liberty*, Blackwell, Oxford, 1946, Chap 3, p. 51.

27 William James, *Psychology*, Henry Holt, New York, 1893, Vol. 1, p. 309; as quoted in Lucius Garvin, *A Modern Introduction to Ethics*, Houghton-Mifflin, Boston, 1953, p. 333.

28 Mill, *op. cit.*, p. 60.

29 John Rawls calls the statement of this psychological tendency "the Aristotelian Principle" and states it as follows: "Other things being equal, human beings enjoy the exercise of their realized capacities (their innate or trained abilities), and this enjoyment increases the more the capacity is realized, or the greater its complexity. The intuitive idea here is that human beings take more pleasure in doing something as they become more proficient at it, and of two activities they do equally well, they prefer the one calling for a larger repertoire of more intricate and subtle discriminations". Rawls cites the preference among good players for chess over checkers and among good mathematicians for algebra over arithmetic. (*A Theory of Justice*, Harvard University Press, Cambridge, Mass., 1971, p. 426).

30 Aptitudes and basic dispositions differ in an important way from ordinary desires, plans and ambitions. The latter characteristically tend to be more precise and determinate than the former, and therefore less flexible and easy to 'fulfill'. Many ambitions are for some relatively specific object and when that object does not come into existence the ambition is denied. General interests, talents, and drives, however, can typically find substitute objects that do equally well. If one has a highly-developed mechanical aptitude, for example, one can employ it equally well as an airplane or an automobile mechanic, as well as a carpenter as a plumber, or in a hundred other callings. One's ambition to be an automobile mechanic, on the other hand, is squelched once and for all, by the denial of opportunity to enter that particular field. For this reason, fulfillment is, on the whole, less difficult to achieve than successful ambition or 'satisfaction'.

31 The distinction between duty and responsibility is well made by J. Roland Pennock in his essay "The Problem of Responsibility" in C. J. Friedrich (ed.), *Nomos III: Responsibility*, Atherton, New York, 1960, p. 13: "We normally reserve [the word *responsibility*] for cases where the performance of duty requires discernment and choice. We might well say to a child, 'It is your responsibility to take care of your room', but we would not be likely to say, 'It is your responsibility to do as you are told' ".

32 Gregory Vlastos, "Justice and Equality", in Richard B. Brandt (ed.), *Social Justice*, Prentice-Hall, Englewood Cliffs, N.J., 1962, p. 44.

33 Thomas Nagel, *op. cit.*, p. 707: "If *sub specie aeternitatis* there is no reason to believe that anything matters, then that doesn't matter either, and we can approach our absurd lives with irony instead of heroism or despair".

34 *Webster's New Collegiate Dictionary* (1976), based on *Webster's Third New International Dictionary of the English Language*. Of the five English dictionaries I consulted, only this newest one contained any definition of irony as an attitude. Is that because this sense is relatively new or because dictionary-makers have heretofore overlooked it?

PETER VAN INWAGEN

PHILOSOPHERS AND THE WORDS 'HUMAN BODY'

> Sing, O Goddess, the ruinous anger of Achilles . . . , which . . . sent too soon to Hades the *psuchai* of many mighty heroes, and made the heroes themselves the food of dogs and carrion birds . . .
>
> *Iliad* 1. 1–5

J. L. Austin once began a paper by listing specimens of sense and specimens of nonsense. I shall do the same.

SPECIMENS OF SENSE

(1) His doctor told him he must not go on abusing his body that way.
(2) Anatomy is the study of the structure of the human body.
(3) Alice told James she hungered for his body.
(4) His body was covered with scars.
(5) The force of the explosion had nearly torn his limbs from his body.
(6) A brain might be removed from one body and implanted in another.
(7) The undertaker laid out Alfred's body.
(8) I believe in the Resurrection of the Body . . .

SPECIMENS OF NONSENSE

(A) A person is [is not] identical with his body.
(B) It is reasonable for each of us to believe that thoughts and feelings are associated with other human bodies.
(C) A person might have [could not have] different bodies at different stages of his career.
(D) I might [could not] have had a different body from the one I have.

Peter van Inwagen (ed.), Time and Cause, 283–299.
Copyright © 1980 *by D. Reidel Publishing Company.*

(E) There might [could not] be two persons inhabiting the same body.

(F) That poor girl in the hospital is really dead, even though the doctors are keeping her body alive.

(G) It is an open question whether, when I use the word *I*, I refer to my body or to some other thing.

(H) After the Resurrection, one will [will not] have the same body one has in the present age.

I

My purposes in this paper are to explain in what sense of 'nonsense' I contend (A)–(H) are nonsense, to argue for the conclusion that (A)–(H) are nonsense, and to explain why my arguments do not lead to the absurd conclusion that (1)–(8) are nonsense.

In what sense do I think (A)–(H) are nonsense? This claim, I admit, is a carelessly stated one. I do *not* mean that the English *sentences* displayed above are 'nonsense'. To call a syntactically correct English sentence made of real English words 'nonsense' must be to claim that that sentence *could not* be used by someone speaking English to express a proposition, ask a question, issue a command, and so on. Or, at least, it must be to claim that that sentence could not be so used unless words were to change their meanings, or unless the user and his audience had established some special convention ("If I say, 'The Absolute is subsumed under a continuum of porous variation', I'll mean 'Orcutt is the spy'"). And I think that at least some among (A)–(H) are not 'nonsense' in this sense. Take (E), for example. This sentence might be used by an English-speaking interplanetary explorer to convey to the other members of his expedition, among whom no special linguistic conventions are in force, a fact about the distribution of population in a certain cluster of planetoids.

So I am not calling *sentences* nonsense. What I mean is roughly this: *typically* when philosophers utter or inscribe sentences like (A)–(H), they are talking or writing nonsense. By *philosophers*, I mean people engaged, perhaps unwittingly, in the philosophical enterprise, whether or not they are professional, academic philosophers. [Someone who utters sentence (F) is almost certainly doing philosophy, even if he has never heard of the subject.] When I say that someone is *talking nonsense* I mean that he is uttering or inscribing words and purports to be expressing a proposition – that is, asserting something or telling how things are – by doing this uttering or inscribing, and that he fails to. (I shall not explicitly discuss nonsensical attempts at questions, commands, and so on. But everything I say will be applicable to uses of language other than assertion, *mutatis mutandis.*) That is to say, to talk nonsense is to represent oneself as saying something when one is only uttering words. But I must append two qualifications to this, or I shall be misunderstood.

First, I have no *theory* about how one should separate sense from nonsense. (By a *theory*, I mean something like the Verification Theory of Meaning.) I do not understand what philosophers say, at least in typical cases, when they utter sentences like (A)–(H), and I think the reason I do not understand them is that they have failed to explain what they mean by the word *body* as they use it. And I think the reason they have failed to explain what they mean is that there is nothing, or nothing coherent, that they do mean. [There are two alternative hypotheses that equally well explain my incomprehension: (i) I do not understand them because I am not clever enough; (ii) I do not understand them because I am too clever by half: I have convinced myself by some sophistical means or other that what is perfectly plain is obscure. Naturally enough, I reject both these hypotheses.]

Second, it is not my intention to insult anyone. It is very hard to make any sense at all when talking about philosophical matters, and it is not necessarily a disgrace for a philosopher to lapse into nonsense. It is not, or *should* not be, any more insulting to say to a philosopher, 'What you say is nonsense', than it is to say, 'What you say is false'. Very great philosophers, philosophers whose shoes I am not fit to untie, have talked mostly nonsense. Descartes's *Meditations*, for example, is almost entirely nonsense. And yet Descartes was a great philosopher, a philosopher deserving of only the very highest admiration. Is this a paradox? Not really. If I know that most of the *Meditations* is nonsense, I did not discover this for myself: I was shown it by a philosopher no less great than Descartes; it was not within *my* power to discover it; the *Meditations* is not *obvious* nonsense. Moreover, nonsense is as variable in its quality as falsity. Newton's theories of motion and gravitation and Lysenko's genetic theories are equally false. Descartes's philosophy and what I wrote in my undergraduate philosophy papers are equally nonsense. But in the same spirit in which one might say of Newton's explanation of planetary motion, 'Would that I were capable of such falsity!', one might quite properly say of Descartes's explanation of voluntary motion, 'Would that I were capable of such nonsense!' (The unwanted air of paradox in what I have just written is aggravated by the fact that one meaning of *nonsense* is 'the obviously and inexcusably false and wrong-headed'. Of course I do not mean the word in that sense.)

Having explained what I am going to argue for, I must now do the arguing. But it is very difficult to show that a certain way of talking is nonsense. The best one can do is examine various explanations of that way of talking and show that they are severally defective. This way of arguing has an obvious weakness: one who employs it can never be sure he has not overlooked some good explanation. I can only say that I shall try not to do this, and concede that someone *might*, for all I say here, give an adequate explanation of what philosophers typically mean by *body*. Besides this obvious weakness, there is a deeper and more subtle weakness in my argument. From the fact that a certain word or idiom cannot be adequately explained, it does not follow that it is meaningless. A case in point is the word *set*. All explanations I have ever seen of the meaning of this word are either *wrong* (e.g., have the consequence that something that is not a set is a 'set') or circular (involve words that are essentially synonyms for *set*). But *set* is as meaningful as a word can be. (We meet other philosophically interesting cases of this phenomenon when we examine proposed definitions of modal and moral terms.) We must not be seduced by what Professor Geach has called 'the Socratic fallacy': for it just *is* possible to use a word correctly – and hence meaningfully – without being able to define it in any nontrivial way.

And yet, *Ich kann nicht anders*. Perhaps if I succeed in showing that various proposed definitions or explanations of *body* are defective, I shall, in the course of doing this, convince others that this word is, in the mouths of philosophers, usually meaningless.

I shall begin by imposing a criterion of adequacy on any definition of *body*: an acceptable definition of *body* must not be *tendentious*. This requirement may be spelled out as follows. Consider those famous and respected philosophers who do think the word *body* (or *corpus*, etc.) is meaningful and who use it in stating their theories. An adequate definition of *body* should not render any of the sentences containing the word *body* that any of these philosophers uses to state his theory a trivial, verbal falsehood. For example, if a Cartesian were to define *body* to mean 'the material thing animated by the immaterial person', this definition would rightly be regarded by the materialist as tendentious. For it renders *his* sentence 'A person is identical with his body' a trivial, verbal falsehood. A way to state essentially the same requirement is this. Consider those philosophers who use the word *body* and who believe that the sentence 'There are bodies' expresses a truth. We must not define *body* in such a way that any of these philosophers could be expected to respond to our definition by saying, 'If *that's* what the word means, then there are no "bodies"'. For example, the materialist might well respond to the 'Cartesian' definition given above by saying something like, 'Since no material things are animated by immaterial persons, there are no such things as "bodies" in *your* sense of the word', and the fact that the materialist could be expected to say this shows that the 'Cartesian' definition is tendentious.

II

Let us begin our search for a definition. The form of words we wish to define is '*x*'s body'. If we understand this form of words then we can define '*x* is a body' in terms of it. Say it some such way as : '*x* is a body' =$_{df.}$ 'There is, was, or will be – or perhaps only "could be"? – a *y* such that *x* is *y*'s body'. If we *begin* by trying to define '*x* is a body', we shall, even if we succeed, still be faced with the task of defining '*x*'s body', and there is no *obvious* way to define '*x*'s body' in terms of '*x* is a body'. (Such *definientia* as 'the *y* such that *y* is a body and *x* animates *y*' and 'the *y* such that *y* is a body and *x* is *y*' are obviously tendentious.) Moreover, we cannot regard it as our sole task to define '*x* is a body' while neglecting '*x*'s body'. For if there is such a concept as the concept of a human body, then simple inspection of philosophers' use of *body* shows that it is part and parcel of understanding this concept to understand what it is for a given body to be the body *of* a given human being.

Let us, for the sake of concreteness, attempt to define 'Descartes's body'. If we discover an adequate definition for 'Descartes's body', then, presumably, our discovery can be generalized in such a way as to provide a definition for '*x*'s body'.

There is a certain sort of answer to our question that is very popular. An excellent example of an answer of this sort is cited in the Oxford English Dictionary:

> 1869 GOULBOURN *Purs. Holiness* ix. 78 By 'the body' is to be understood the mass of matter which we carry about with us, with all the animal properties that belong to it.

(An erudite historian of philosophy – not, I think, so erudite as to have read Goulbourn – once gave me an almost verbally identical definition of *body* when I pressed him for one in the course of a philosophical discussion.) This definition suggests the following definition of 'Descartes's body':

> Descartes's body $=_{df}$. The mass of matter Descartes carried about with him, with all the animal properties that belonged to it.

But this is very puzzling. I doubt very much whether there was a mass of matter – as opposed to a mass of . . . ? – that Descartes carried about with him *always*; and if there was, I should expect it to have been something like a good-luck charm, which would presumably have been devoid of 'animal properties'. But suppose Descartes did always carry about with him something possessing animal properties: suppose he always carried about a *mouse*, like Benjamin Franklin in the children's book *Ben and Me*. Then, so far as I can see, this mouse would have satisfied our definition and would have been 'Descartes's body'.

I am not being flippant. And if I am being deliberately obtuse, it is with what I conceive as a good purpose. I really *am* aware that Goulbourn's words "carry about with" are supposed to be understood metaphorically. But what is the 'cash value' of the metaphor? I do not know. I think no one knows.

Perhaps I am being too hard on poor Goulbourn. Perhaps he expressed himself carelessly and attempted to convey a simple and straightforward idea by unsuitable means. I can imagine a defender of Goulbourn saying impatiently, 'Look, this is mere pedantry. It's easy enough to see and to express clearly what Goulbourn meant to say:

> Descartes's body $=_{df}$. The mass of matter (i.e., the material thing) that was the *bearer* of all Descartes's animal (and, more generally, physical) properties.'

But this will not do. For it was Descartes himself who was the bearer of all Descartes's animal and physical properties. That is, Descartes had such animal and physical properties as Descartes had. This is a simple logical truth, a truth that must be accepted by anyone, whatever his metaphysical opinions may be. Therefore, it is a trivial, verbal consequence of our definition that either Descartes was identical with his body (if he had some animal properties), or else there was no such thing as Descartes's body (if he had no animal properties). But this disjunctive proposition is incompatible with dualism, and hence our definition must be rejected as tendentious.

I can, I fear, imagine a defender of our imaginary defender of Goulbourn saying (of course, impatiently), 'Look, this is mere pedantry. It's easy enough to see and to express clearly what Goulbourn's defender meant to say:

> Descartes's body $=_{df}$. The material thing that bore the animal (or physical) properties commonly if loosely predicated of Descartes.'

But what does 'commonly if loosely predicated of Descartes' mean? Whatever precisely this condition may come to, surely it is satisfied by any property Descartes was commonly believed to possess. Now suppose Descartes was commonly believed to suffer from kidney stones (*being a sufferer from kidney stones* is surely an 'animal property'), though this common belief was false. Then there was *nothing* that bore all the animal properties commonly if loosely predicated of Descartes, and there was no such thing as Descartes's body. Or suppose Descartes had a twin brother who shared all the animal properties Descartes was commonly believed to have. Then nothing uniquely satisfied the condition laid down in our definition, and, again, there was no such thing as Descartes's body. This second example, by the way, would also serve to refute any attempt to emend our definition by laying it down that Descartes's body was the material thing that bore *most of* the animal properties commonly predicated of Descartes.

III

Let us abandon our attempts to repair Goulbourn's definition and look at a different sort of attempt to show what *body* means. In this section and the following sections, I shall imagine people saying various things to me: offering definitions of *body*, qualifying them, objecting to my replies, and so on. These imaginary conversations are based on real conversations I have had.

A philosopher points at me and says, '*This* is your body. It may *also* be *you* – whether it is is an open philosophical question – but it is what I call "your body".' (Of course, a similar definition could have been given for 'Descartes's body'. In this section I shall try to understand 'my body' rather than 'Descartes's body', since Descartes is unfortunately not available to be pointed at. The fact that I shall be trying to understand '*my* body' rather than '*his* body' has no particular significance.)

Well, *what* does he point at? (I said he points at *me*, but perhaps I ought not to have said that. At any rate, you can imagine what his gesture is supposed to have looked like.) He might have used the same gesture, the same tone of voice, the same glance, in referring to my skin or my navel or my abdomen. But these are not what he was pointing at, though an observer sighting along my philosopher's index finger as along the barrel of a gun would have found all these things or parts of them at the center of his field of vision. When someone points and says, '*That* is such-and-such', we may always ask, 'That *what*?', and the answer we receive ('Man', 'Vase', 'Number',

'Grass', 'Star', 'Empty space', 'Country', 'Sentence', 'Handwriting') tells us what was being pointed at.

So I ask, 'That *what*?' What will my philosopher say? He cannot say, 'That body', for this will not help me. He had better not say, 'That man' or 'That human being' for *I* am a man and a human being, and if he points at a human being, he points at *me*. (Well, I might turn out to be a Martian foundling or a robot, but in such a case, there would not be any man or human being there for him to point at. Insofar as it is controversial whether I am a human being, it is controversial whether any human being is where I am.) Or perhaps this is wrong. One might interpret Descartes, and perhaps Locke, as saying this: *I* am not a human being, for I am a soul, and a human being is a sort of compound of a soul and a body. Well, of course, I do not understand this view, because I do not understand *body* – or *soul* – in this context. Moreover, my respondent himself must admit that if this view is correct, then he cannot ostensively refer to my body as 'this human being'. Of course he might call it 'the body of this human being', but this latter phrase would serve his purpose no better than 'this body'.

Perhaps he will answer, 'This material object'. But what does he mean by this? The words 'material object' are a philosophers' term of art. They constitute a 'stand-in phrase' whose function is to save a philosopher's having to say 'vase, ashtray, teacup, stone, . . . ', every time he wants to state a general philosophical thesis. (Of course, various philosophers have, for their own several purposes, *defined* 'material object'. But the only generally agreed-upon condition on the adequacy of such a definition is that 'moderate-sized specimens of dry goods' must count as material objects.) There are many things such that it is not clear whether they are 'material objects'. The ashtray is a material object, but what about the left half of the unbroken ashtray? What about the ashtray and the teacup 'taken together'? What about a cloud? A galaxy? A proton?

More to the point, what about men and women? If a man is uncontroversially a material object – that is, if *man* is one of those words that, as a matter of the *meaning* of 'material object', should occur on a complete list of words that 'material object' can properly stand in for – then my philosopher has pointed at me, for I am a man, whatever Descartes and Locke may say: a certain man is the only 'material object' *there*. On the other hand, if the meaning of 'material object' precludes its standing in for a list of words including *man*, then there is no material object for my philosopher to point at: a certain man occupies the region of space he is indicating, and there is no *room* for any 'material object'.

A third possibility is this: the meaning of 'material object' is such as to leave it an open question whether these words can stand in for a list of words that includes *man*. But in that case, it seems to me, it is an open question whether my philosopher succeeds in pointing at *anything*. There is a man before him, after all. If it is not clear whether he is pointing at that man, then it is not clear what he *is* pointing at.

These difficulties arise only if 'material object' is simply a 'stand-in phrase'. Let us therefore explore the consequences of *defining* 'material object'. Let us say that a 'material object' is a *ponderable* object: one that weighs something. To achieve greater generality of application, we might say that a material object is an object that has a mass, but we need not be so subtle. Other definitions – e.g., 'object that occupies space' – might be examined, but I think they would all have consequences like those I shall argue follow from the definition we are considering.

My philosopher says, '*This* is your body'. I ask, 'This *what*?' He replies, 'This thing that weighs something'. But I weigh something. Rather too much, in fact. Is he referring to me? I do not see how to avoid this conclusion. Some philosophers think they do. They point, for instance, to the fact that a truck driver may say to his dispatcher, 'I weigh seven tons', when, strictly speaking, he weighs far less. But the truck driver *could* say, 'My truck weighs seven tons' instead of 'I weigh seven tons'. What could I say in place of 'I weigh something'? Perhaps, 'My body weighs something'? But this I do not understand.

Compare this case. A certain philosopher holds that nothing has both color and shape. I say that I should have thought that this saucer was both blue and round. He replies that it is the *morphosome* of the saucer that is round and the *chromosome* of the saucer that is blue. I say that I do not understand what he means by, for example, the 'morphosome' of the saucer. He points at the saucer – or so *I* should describe his gesture – and says, '*This* is the morphosome of the saucer'. I say, 'You *seem* to be pointing at the saucer. You haven't yet got me to see what you mean by the morphosome of the saucer'.

This little story suggests a reply that my philosopher (the one who is trying to define *body* by ostension) may make to my objections. He may say, 'You're just a materialist. You think a person is identical with his body. But I have got you to see what I *mean* by *body*. Your insistence that I'm pointing at *you* derives from the meaning I have provided for *body plus* your theory, and not from that meaning alone'. We might compare this with, 'You're just a morphosomatist. You think a saucer is identical with its morphosome. But

I *have* got you to see what I *mean* by *morphosome*. Your insistence that I'm pointing at *the saucer* derives from the meaning I have provided for *morphosome plus* your theory, and not from that meaning alone'. But I *do not* think either that a man is identical with his body or that a saucer is identical with its morphosome. If a man does what I would call 'pointing at a saucer', or performs an act of pointing I cannot distinguish from pointing at a saucer, and says, 'That's the saucer's morphosome', I am not thereby enabled to understand *morphosome*. If a man does what I would call 'pointing at me', or performs an act of pointing I cannot distinguish from pointing at me, and says, 'That's your body', I am not thereby enabled to understand *body*. And since I do not understand *body*, I do not understand the accusation that I think a man and his body are one and the same.

IV

Let us try something else. 'You know what your arms are, don't you? And your hands, trunk, legs, feet, neck, and head? Well, all these things taken together are your body. Your body may or may not be *you*; that's an open question." This explanation *seems* to imply that many things and one thing are somehow identical, this logical absurdity (I had to employ a piece of grammatical sleight of hand even to get it stated without falling into obvious incoherence) being disguised with the help of the useful phrase 'taken together'. But I think all that is meant is that the many things are *parts* and are all the parts, other than their own parts, of the one. At least this is the only sense I can make of it. This suggests the following definition (we may now return to Descartes, since we no longer have to employ ostension in our definition):

> Descartes's body $=_{\text{df}}$ The thing such that Descartes's limbs, trunk, and head [I abbreviate the above list thus] were parts of it, and everything that was part of it that was not one of these things was composed of parts of these things.

I have certain difficulties in understanding this definition. Suppose that Descartes's left hand was surgically removed, in such a way as to leave him otherwise unimpaired, on the first day of 1625. Suppose it was frozen solid throughout 1625, and burned to ashes on the first day of 1626. Taken literally, our definition seems to entail the following:

> In 1624, Descartes's body was all in one piece. Throughout 1625, it existed in a spatially separated state (i.e., it was not 'in one piece'). In 1626, it ceased to exist, though Descartes continued to exist.

My impression is that these consequences will not be welcomed by most people who use *body*. Moreover, I do not wish to form a definition that legislates about questions of 'mereological essentialism'. Let us say, therefore, what it is for something to be Descartes's body at a particular moment. This will leave it an open question whether, if one of Descartes's members ceased to exist at *t*, the body, if any, Descartes had after *t* was the same body he had before *t*.

> The thing that is Descartes's body at $t =_{df}$. The thing such that, for any *x*, if *x* is one of Descartes's limbs at *t*, or is his trunk at *t*, or is his head at *t*, then *x* is part of it at *t*, and, for every *y*, if *y* is part of it at *t*, and *y* is not one of Descartes's limbs at *t* and not Descartes's trunk at *t* and not Descartes's head at *t*, then *y* is composed at *t* of parts of the things that are his limbs, trunk, and head at *t*.

If we are not careful about the meaning of *part*, we shall find that this definition has a very strange consequence. Suppose that Descartes was burned at the stake and that only the finest ashes remained: nothing that existed thereafter was a limb of Descartes or Descartes's trunk or Descartes's head. Let *T* be a moment of time later than this *auto-da-fè* and suppose that at *T* there existed exactly one thing without parts – a Democritean atom, say. It is easy to see that, according to our definition, this atom was Descartes's body at *T* (and doubtless the body of a good many other people). But we can avoid this consequence if we assume that everything is a part of itself, that is, if we stipulate that a 'part' of an object need not be a *proper* part of that object. Then nothing is without parts, not even a Democritean atom, and the absurd conclusion we drew in the last sentence but one can be seen not to follow from our definition.

Still, this definition is not without its odd consequences. Suppose someone had severed one of Descartes's limbs – his left leg, say – and had destroyed his remaining limbs and his trunk and his head. Let *T* be any moment of time thereafter, provided only that Descartes's severed left leg still existed at *T*. Then, according to our definition, this severed left leg was Descartes's body at *T*. (People who think they understand *body* tell me that any definition that has this consequence is wrong.) Moreover, suppose some marvelously advanced seventeenth-century surgeon had succeeded in grafting the severed leg of Descartes on to the stump of the leg of some contemporary of Descartes's who had lost his leg in an accident. Then, according to our definition, this man could have said truly, 'Descartes's body is part of my body'.

But perhaps this odd consequence does not follow. Someone may object, 'Your argument assumes that when Descartes's leg was severed, it remained Descartes's leg. But this is not literally true. We *talk* that way, of course, but we are then only identifying the limb by reference to its *former* owner, like country people who call a farm "the Harkness place" when no Harkness has lived there for forty years. After Descartes's leg was cut off, it was, strictly speaking, *no one's* leg. If it was then grafted on to someone's stump and its tissues sustained by the circulation of that man's blood, then it *became* that second man's leg. It's really very like the transference of a piece of property from one owner to another'.

But this response raises certain difficulties, or, better, makes evident difficulties that have been present all along. These difficulties arise when one attempts to answer the question, 'What *does* it mean to say that a certain leg was, at a certain moment, "Descartes's"?' One might, of course, say that a certain leg was Descartes's (at *t*) if it was part of Descartes's body (at *t*). But, of course, if this is what 'Descartes's leg' *means*, then 'Descartes's leg' cannot occur in a definition of 'Descartes's body'.

Now *I* should say that 'Descartes's leg' means 'leg that was part of Descartes'. If I am right about this, then the definition we are considering is tendentious. *If* 'Descartes's leg' *means* 'leg that was part of Descartes', then it is a trivial, verbal consequence of the present definition that Descartes was identical with his body. Or, more precisely, it is a trivial, verbal consequence of this definition that at any moment *t*, Descartes was identical with the thing that was his body at *t*.

I can see nothing that 'Descartes's leg' could mean besides 'leg that was part of Descartes's body' and 'leg that was part of Descartes'. If it means either of these, the present definition is defective. Therefore, the present definition is defective.

V

Let's try again: 'You know what a *dead* body is, don't you? It's what you see laid out at the undertaker's, and what goes into the grave. Well, the dead body you see was once alive, but, alive or dead, what you see is and was a body'.

But I should say that what I see laid out at the undertaker's is a dead *man* (or woman or child). The man I see lying on the bier was once alive and is now dead, just as the man I see lying in the hospital bed was once healthy and is now ill. (Actually, this is only a first approximation of what I want to say about corpses. I shall presently introduce a qualification.) Descartes, a few moments before his death, could truly have said, 'In a few moments, I (this thing that thinks) shall be a corpse', though doubtless, given his philosophical views, he *would not* have said this. And each of us, unless he is fated to suffer a particularly violent death, will one day be a corpse.

'But that can't be right, for when you die you cease to exist. But, at least in most cases, your corpse exists after you die. So it can't be true that we shall one day be corpses.' This objection can hardly win the assent of *everyone* who thinks that *body* is a meaningful word. Descartes, for example, would not have agreed that he should cease to exist after his death. Moreover, it is not true. In *some* cases a person ceases to exist at the instant he dies: the man who makes a wrong move when disarming a powerful bomb, for example. But this is the exception rather than the rule. In the normal case, even supposing that there is no 'afterlife', one exists after one dies. I hasten to add that I see no particular advantage in being a corpse over not existing at all; and neither of these is a more or less enviable state than being alive but in a perpetual and dreamless sleep.

Here is the qualification I promised: a corpse is not always a dead man or woman or child. A fresh, well-preserved corpse is, but what we see at the undertaker's is not a fresh, well-preserved corpse but something like the shell or ruin of a dead man. I use the words *shell* and *ruin* by analogy to the way these terms are used to talk about buildings. If a building has been gutted by fire or very nearly destroyed in some other way, and if the fire, or whatever, has left standing something that suggests the original outward appearance of the building, then we call what still stands the 'shell' or 'ruin' of the building. Now think for a moment about what the embalmer has done to the corpse you see at the undertaker's, and I think you will see the appropriateness of calling what you see there the 'shell' or 'ruin' of a man, rather than the man himself. So I do not claim that, for example, Lenin can be observed to this day in a glass case in Moscow: what lies in the case is not, literally speaking, Lenin, but rather his shell or ruin; his 'remains', as they say.

What the point of transition is between what is literally a dead *man* and what is a mere ruin of the man that was, I should not like to say.

VI

> For should the Soul of a Prince, carrying with it the consciousness of the Prince's past Life, enter and inform the Body of a Cobler as soon as deserted by his own Soul, every one sees, he would be the same person with the Prince, accountable only for the Prince's Actions . . .
>
> Locke, *Essay* 2.27

'Bodily transfer' – well, if such stories make sense, then, obviously, *body* makes sense. But, of course, I do not understand Locke's story. 'But I can imagine evidence', says someone, 'that would force you to concede that the soul of a prince had entered and informed the body of a cobbler.' I doubt it, just as I doubt anyone could imagine evidence that would force me to concede that the Absolute issues both imperatives and indicatives. But we all know how the Prince-and-Cobbler story goes, and I will not bore you with it. Let us call the cobbler 'Harry' and the prince 'Charlie'. 'How would you explain the fact that Harry, or someone who looks just like Harry and is in Harry's house, exhibits both an unshakable conviction that he is Charlie and perfect knowledge of the most particular and intimate facts of Charlie's history?' Well, frankly, if I were confronted with such a case, I should not know *what* to say. And neither would you or anyone else *know* what to say. But I can think of many consistent stories that would explain the observable facts of the case, though it would probably be hard to find out which of these stories was true. Here is one. The Martians (who, as Norman Malcolm once observed, are wonderfully technologically advanced in the imaginations of philosophers) perform something like plastic surgery – but much more complete and effective – on Charlie, thereby causing him to look just like Harry; they remove Harry and, faster than the eye can follow, replace him with the altered Charlie. 'But this would be pure fantasy. Wouldn't it be simpler to assume that Charlie's soul, or mind, or personality, or *persona*, had somehow entered Harry's body?' Of course it would be pure fantasy. Is that not appropriate? The Harry-and-Charlie story is pure fantasy to start with. If the Empire State Building grew wings and flew away, and everyone agreed this really had happened, could we call *any* proposed explanation of this event 'just too fantastic'? If a cobbler started talking like a prince (or however we should describe the purely visual and auditory component of the

Harry-and-Charlie story), then *something* pretty odd would be going on in the world.

Let *me* propose a case. Suppose Harry had a twin brother (a plumber), and they *both* started acting like Charlie. What could be said about this case? Note that it is not a *more* fantastic case than the first. A story in which both the Empire State Building and the Chrysler Building fly away is not *more* fantastic than a story in which the Empire State Building alone departs in this remarkable fashion.

VII

I think I had better bring this essay to a close. I have not examined every possible attempt to make sense of *body*. Other things might be done. For example, one might attempt to combine the methods used in sections III and IV, avoiding the problem of defining 'Descartes's leg' by using in its place, 'this leg'. But arguments ideally have two sides and perhaps it is time for the other side to say something. I have been examining possible defenses of a certain view – that *body* is meaningful – and attempting to refute these defenses. Perhaps those who actually hold this view can construct better defenses of it than someone who, like me, rejects it. All other things being equal, the most effective attorney for the defense is the one who believes in the innocence of his client.

In this section, I shall try to say what I think the difference is between my 'specimens of sense' and my 'specimens of nonsense'. That is, I shall try to explain how it is that, if *body*, as philosophers typically use this word, has no meaning, ordinary speakers can utter any of the 'specimens of sense' and succeed in making an intelligible statement. Roughly speaking, I should say that this is the case because *body* occurs *essentially* in (A)–(H), and only *accidentally* in (1)–(8). That is to say, (1)–(8) can be paraphrased into sentences in which *body* does not occur, and no such paraphrases can be given for (A)–(H). Consider these paraphrases of (1)–(8):

> His doctor told him he must not go on abusing his health in this way.
> Anatomy is the study of the physical structure of human beings.
> Alice told James she wanted to have sexual intercourse with him.
> He was covered with scars.
> The force of the explosion had nearly torn off his limbs.
> It might be possible to remove the brains from two people and implant one in the space formerly occupied by the other.

The undertaker laid out Alfred's corpse.
I believe that one day God will restore the dead to life.

I would say that each item on this list means *approximately* what the corresponding item on the list of specimens of sense means. I could make the match exact if I were willing to write longer paraphrases, but I wish only to indicate roughly how such paraphrases should go. I think it is evident that the 'specimens of nonsense' cannot be paraphrased in this way. If any philosopher thinks he *can* restate any of (A)–(H) in such a way that the restatement does not involve *body* or some synonym, I shall be willing to look at his paraphrases. But paraphrases, even ones I can understand, will not help me much unless all, or at least, most, philosophers who use *body* will agree that these paraphrases capture the meaning of the original specimens. It is (A)–(H) *as typically used* that I want to understand, after all. Anyone can propose *some meaning or other* for *body*; but if the generality of philosophers insist that this proposed meaning is not *theirs*, then I shall not have been helped to understand *them*.

There are at least two important questions that are closely related to the topics discussed in this essay. The first is: What are the implications of what has been said in this paper for the traditional problems of philosophical psychology and metaphysics (e.g., for the problem of personal identity)?

The second question is this: What is the origin of the idea of a 'human body' – that is, what is the origin of the way we talk when we use these words – and why does this idea have such a grip on us?

I have no space here to say anything adequate in response to the first question. Perhaps I shall return to it in another essay.

The second question is at least partly a question that belongs to the history of ideas, and I, who am no profound scholar, am perhaps not the best person to attempt to answer it. I will venture a guess, however, that 'the body' is an invention of philosophers and theologians. I think it is quite wrong to suppose that primitive, prephilosophical man had any such idea. Some people, I learn from reading works on archaeology, think that gifts and tools placed in ancient graves with evidently elaborate ritual, show that primitive man believed that at death a man's spirit departs to dwell in another place, leaving his body behind. And, of course, what can be *left behind* is something that was *there* all along. Well, I do not dispute the graves and the tools and the gifts and the ritual. But I doubt whether we can infer from them that the gravediggers were Platonic dualists, or even that they held a perhaps confused view that could have been turned into Platonic dualism by a little Socratic dialectic.

Consider Homer's audience. As the epigraph of this essay shows, it was obvious to these men – and it *ought to* be obvious to you and me – that if a man is killed in battle, then it is *he* who lies dead on the stricken field, soon to be burned or to rot or to be eaten by dogs. In the Homeric view, something of the slain man does indeed depart to dwell in another place: his *psuche*, that is, his *breath* or his *life*. (*Cf*. 'No sooner was the breath out of his body . . . '; 'No sooner had life left him . . . '.) But the *psuche* is not the man himself. Indeed, in an admittedly extraordinary case, the *psuche* of a man is in Hades though he himself is alive and elsewhere (Heracles in *Odyssey* 11). It is a measure of the strength of the grip that the idea of 'the body' has on us that almost any edition of the *Iliad* you may care to examine will contain some such note as this: "I, 4 *αὐτοὺς*: *themselves*; i.e. their bodies". Many editors compound this confusion by adding something along the lines of, "For Homer, a person is identical with his body", as if Homer were a modern philosophical materialist like David Lewis!

Burial customs may indeed show that primitive man had some sort of idea of an 'afterlife'; the example of Homer, however, shows us that a belief in an afterlife does not imply a belief in 'the body'. Therefore, I rest easy in my contention that the body, like the unreality of motion, the bare particular, and the Absolute Idea, is an invention of philosophers. I think it is a bad invention. I invite you to consider carefully the possibility of discarding it.

Syracuse University

A BIBLIOGRAPHY OF THE PHILOSOPHICAL WRITINGS OF RICHARD TAYLOR

1950

'Comments on a Mechanistic Conception of Purposefulness', *Philosophy of Science* **17**. Reprinted in *Purpose in Nature*, ed. by J. V. Canfield (Englewood Cliffs, N.J., 1966) and in *Modern Systems Research for the Behavioral Scientist*, ed. by Walter Buckley (Chicago, 1968) and in *Philosophy, the Cutting Edge*, ed. by David Berlinski (Sherman Oaks, Calif., 1976).

'Behavior and Purpose: A Rejoinder', *Philosophy of Science* **17**. Reprinted in Buckley, *ibid.*

1951

Review of *A History of Ancient and Medieval Philosophy* by Frederick Mayer. *Philosophy and Phenomenological Research* **11**.

1952

'Mr Black on Temporal Paradoxes', *Analysis* **13**.

'Negative Things', *The Journal of Philosophy* **49**.

'Mr Wisdom on Temporal Paradoxes', *Analysis* **13**.

'The Writings of C. J. Ducasse' [a bibliography], *Philosophy and Phenomenological Research* **13**.

Review of *Theory of Knowledge* by A. D. Woozley. *The Philosophical Review* **61**.

Review of *St. Thomas Aquinas on Aristotle's Love and Friendship* by Pierre Conway. *Philosophy and Phenomenological Research* **12**.

Review of *Introduction to the Philosophy of Nature* by R. A. Kocourek. *Philosophy and Phenomenological Research* **12**.

Review of *The Doctrine of Being in the Aristotelian Metaphysics* by Joseph Owens. *Philosophy and Phenomenological Research* **13**.

1953

'Ayer's Analysis of Negation', *Philosophical Studies* **4**.

'A Note on Knowledge and Belief', *Analysis* **13**.

'A Commentary on Aristotle's Ethics', *Review of Metaphysics* **7**.

Review of *A Study in Memory* by E. J. Furlong, *The Journal of Philosophy* **50**.

1954

'Rejoinder to Mr Malcolm', *Analysis* **14**.

'Knowledge and Certainty', *Review of Metaphysics* **7**.
'Disputes about Synonymy', *The Philosophical Review* **63**.

1955

'Spatial and Temporal Analogies and the Concept of Identity', *The Journal of Philosophy* **52**. Reprinted in *Problems of Space and Time*, ed. by J. J. C. Smart (New York, 1964).
Review of *Sovereign Reason* by Ernest Nagel, *U.S. Quarterly Book Review* **11**.

1956

'Knowing What One Knows', *Analysis* **16**.
'The "Justification" of Memories and the Analogy of Vision', *The Philosophical Review* **65**.
'Does it Make Sense to Suppose that All Events, Including Personal Experiences, Could Occur in Reverse?', *Analysis* **16**.

1957

'The Problem of Future Contingencies', *The Philosophical Review* **66** (lead article). Reprinted in *Essays in Ancient Greek Philosophy*, ed. by John P. Anton (Albany, 1970).
Theism by J. S. Mill. Edited and with a new introduction by R. Taylor (The Liberal Arts Press, New York).

1958

'Determinism and the Theory of Agency', in *Determinism and Freedom*, ed. by Sydney Hook (New York University Press, New York).
'On Seeing Double', *The Philosophical Quarterly* **8**. [With Timothy Duggan]

1959

'Moving About in Time', *The Philosophical Quarterly* **9**.

1960

'I Can', *The Philosophical Review* **69**. Reprinted in *Free Will*, ed. by S. Morgenbesser and J. Walsh (Englewood Cliffs, N.J., 1962) and in *Readings, in Philosophical Analysis*, ed. by H. Feigl, W. Sellars, and K. Lehrer (New York, 1973) and in *Introduction to Philosophy: A Contemporary Perspective*, ed. by R. Hoffman and S. Gendin (Belmont, Calif., 1975).
'Making Things to Have Happened', *Analysis* **20**. [With R. M. Chisholm]
'Pure Becoming', *The Australasian Journal of Philosophy* **38**.

1961

'Faith', in *Religious Experience and Truth*, ed. by Sydney Hook (New York University

Press, New York). Reprinted in *Philosophy of Religion*, ed. by S. Cahn (New York, 1971).

Review of *The Scottish Philosophy of Common Sense* by S. A. Grave. *The Philosophical Review* **70**.

The Will to Live: Selected Essays of Arthur Schopenhauer, ed. and with an introduction by R. Taylor (Doubleday and Co., New York).

An Essay Concerning Human Understanding by John Locke, ed. and abridged by Richard Taylor. Included in *The Empiricists* (Doubleday and Co., New York).

1962

'Fatalism', *The Philosophical Review* **71**. Reprinted in *The Philosophy of Time*, ed. by R. Gale (New York, 1967).

'Fatalism and Ability', *Analysis* **23**.

Review of *Rationalism in Greek Philosophy* by George Boas. *Brown Alumni Monthly* **62**.

1963

Metaphysics (Prentice Hall, Inc., Englewood Cliffs, N.J.), pp. 85–94, reprinted as 'God' in *Introduction to Philosophy: A Contemporary Perspective*, ed. by Robert Hoffman and Sidney Gendin (Belmont, Calif., 1975), and as 'A Reformulation of the Argument From Contingency', in *God, Man and Religion*, ed. by Keith E. Yandell (New York, 1973), and as 'The Cosmological Argument', in *The Problem of Philosophy*, ed. by C. F. Delaney, Gary Gutting, Michael J. Loux and Harold Moore (Notre Dame, Indiana, 1976). Pp. 96–102 reprinted in *Problems and Perspectives in the Philosophy of Religion*, ed. by George O. Mavrodes and Stuart C. Hackett (Boston, 1967), and as 'A Defense of Theism', in *A New Introduction to Philosophy*, ed. by Steven M. Cahn (New York, 1971), and as 'A Contemporary Form of the Design Argument', in *The Art of Philosophy*, ed. by Fred Westphal (Englewood Cliffs, N.J., 1972).

'The Stream of Thoughts vs Mental Acts', *The Philosophical Quarterly* **13**.

'Causation', *The Monist* **47**. Reprinted in *The Nature of Causation*, ed. by M. Brand (The University of Illinois Press, Urbana).

'A Note on Fatalism', *The Philosophical Review* **72**.

Articles LEWIS, CLARENCE IRVING and ROYCE, JOSIAH in the *Encyclopedia International* (The Grolier Co., New York).

1964

'Deliberation and Foreknowledge', *American Philosophical Quarterly* **1**. Reprinted in *Free Will and Determinism*, ed. by B. Berofsky (New York, 1966) and in *Management: A Decision Making Approach*, ed. by S. Young (Belmont, Calif., 1968) and in *Introduction to Moral Philosophy*, ed. by P. Davis (Columbus, 1972).

'Can a Cause Precede its Effect?', *The Monist* **48**.

'Not Trying to do the Impossible', *The Australasian Journal of Philosophy* **42**.

'Fatalistic Arguments: Comment', *The Journal of Philosophy* **61**.

'Arthur Schopenhauer', in *A Critical History of Western Philosophy*, ed. by D. J. O'Connor (Collier-Macmillan, Ltd., London).

'Justice and the Common Good', in *Law and Philosophy*, ed. by S. Hook (New York University Press, New York).

Review of *Schopenhauer* by Patrick Gardiner, *The Journal of Philosophy* **61**.

1965

'Time, Truth, and Modalities', *Mind* **74** [with Keith Lehrer].

'Time, Truth, and Ability', *Analysis* **25** [with Steven Cahn, under the pseudonym Diodorus Cronus].

Introduction to *The Ontological Argument*, ed. by Alvin Plantinga (Doubleday and Co., New York).

Introduction to *The Basis of Morality* by A. Schopenhauer, translated by E. F. J. Payne (The Bobbs-Merrill Co., New York).

Metafisica, Spanish translation of *Metaphysics*, by Carlos Gerhard (Union Tipografica Editorial Hispano-Americano, Mexico City).

Metafysica, Dutch translation of *Metaphysics*, by A. M. Offermans-van Galen (Prisma Boeken, Utrecht).

1966

Action and Purpose (Prentice-Hall, Inc., Englewood Cliffs, N.J.). Pp. 57–72 reprinted as 'Simple Action and Volition', in *The Nature of Human Action*, ed. by Myles Brand (Glenview, Ill., 1970). Pp. 35–39 reprinted as 'The Metaphysics of Causation', in *Causation and Conditionals*, ed. by Ernest Sosa (Oxford, 1975). *Action and Purpose* has been reprinted by Humanities Press (New York, 1974).

'Prevention, Postvention, and the Will', in *Freedom and Determinism*, ed. by Keith Lehrer (Random House, New York).

1967

'Shadow-Boxing Critics', *Pacific Philosophy Forum* **6**.

Articles CAUSATION, DETERMINISM, and VOLUNTARISM in *The Encyclopedia of Philosophy*, ed. by Paul Edwards (The Macmillan Co., New York and London).

Review of *Cause and Effect*, ed. by Daniel Lerner, *The Philosophical Review* **76**.

The Will to Live: Selected Essays of Arthur Schopenhauer, ed. and with an introduction by R. Taylor (new edition) (Frederick Ungar Publ. Co., New York).

1968

'Dare to be Wise', *Review of Metaphysics* **22**. Reprinted in *Preface to Philosophy*, ed. by James A. Gould (New York, 1969) and in *Teaching Philosophy Today*, ed. by T. W. Bynum and Sydney Reisberg (Bowling Green, Ohio, 1978).

'Law and Morality', New York University *Law Review* **43**.

Japanese translation of *Metaphysics*, Tokyo.

1969

'How to Bury the Mind-Body Problem', *American Philosophical Quarterly* **6**. Reprinted in *Exploring Philosophy*, ed. by Peter A. French (Cambridge, Mass., 1970) and in *Philosophy and Contemporary Issues*, ed. by John R. Burr and Milton Goldinger (New York, 1976).

'Deliberation and Freedom', *Southern Journal of Philosophy* **6**.

'The Anatta Doctrine of Personal Identity', *Philosophy East and West* **19**.

'Thought and Purpose', *Inquiry* **12**. Reprinted in *The Nature of Human Action*, ed. by Myles Brand (Glenview, Ill., 1970).

Metafísica, Portugese translation of *Metaphysics* by Alvaro Cabral, Rio de Janeiro.

1970

Good and Evil: A New Direction (The Macmillan Co., New York). Ch. 18, 'The Meaning of Life', reprinted in *Philosophical Issues*, ed. by James Rachels and Frank A. Tillman (New York, 1972) and in *Coming of Age in Philosophy*, ed. by Roger Eastman (San Francisco, 1973) and in *Ethical Issues in Death and Dying*, ed. by T. L. Beauchamp and S. Perlin (Englewood Cliffs, N.J., 1978) and in *Occasions for Philosophy*, ed. by James C. Edwards and Douglas MacDonald (Englewood Cliffs, N.J., 1979). Pp. 229–235 reprinted as '*Eros*, or the Love of the Sexes', in *Social Ethics*, ed. by Thomas A. Mappes and Jane S. Zembaty (New York, 1977).

'Can There be a Science of Human Nature?', *Proceedings of the International Philosophy Year* **2**. Reprinted in *Mind, Science and History*, Vol. 2 of *Contemporary Philosophic Thought*, ed. by Howard A. Kiefer (Albany, 1970).

'Response to Professor A. J. Ayer', *Proceedings of the Center for Philosophic Exchange*, Summer.

Review of *Action* by Sir Malcolm Knox. *The Philosophical Quarterly* **20**.

1971

'The Governance of the Kingdom of Darkness', *Southern Journal of Philosophy* **9** [under the pseudonym Diodorus Cronus]. Reprinted in *Philosophical Explorations*, ed. by Peter A. French (Morristown, N.J., 1975).

1972

'Ha'oz Leheyoth Chekem', *Iyyun: Hebrew Philosophical Quarterly* **22** [Hebrew translation of 'Dare to Be Wise'].

Preface to *Motive and Intention* by Roy Lawrence (Northwestern University Press, Evanston, Ill.).

1973

Freedom, Anarchy, and the Law (Prentice Hall, Inc., Englewood Cliffs, N.J.).

With Heart and Mind: A Picture (St. Martin's Press, New York).

'De Anima', *American Philosophical Quarterly* **10**. Reprinted in *Language, Metaphysics, and Death*, ed. by John Donnelly (New York, 1979).

'Donnelly on Good and Evil', *International Philosophical Quarterly* **13**.

1974

Metaphysics, Second edition, revised and enlarged (Prentice-Hall, Inc., Englewood Cliffs, N.J.), Ch. 5, 'Freedom and Determinism', reprinted in *Reason and Responsibility*, 4th ed., ed. by Joel Feinberg (Encino, Calif. and Belmont, Calif., 1978). Pp. 30–37 reprinted as 'Reality Consists of Matter', in *Classic Philosophical Questions*, 3rd ed., ed. by James A. Gould (Columbus, 1979).

Introduction to *The Fourfold Root of the Principle of Sufficient Reason* by A. Schopenhauer, translated by E. F. J. Payne (The Open Court, Chicago).

1975

'A Tribute' [to R. M. Chisholm], in *Analysis and Metaphysics*, ed. by Keith Lehrer (D. Reidel, Dordrecht).

1976

'What Do Philosophers Know?', in *Introductory Philosophy*, ed. by Matthew Lipman (Prentice Hall, Englewood Cliffs, N.J.).

'Socratic Ethics and Justice', in *Moral Philosophy, an Introduction*, ed. by Jack Glickman (St. Martin's Press, New York).

'Action and Responsibility', in *Action Theory*, ed. by Myles Brand and Douglas Walton (D. Reidel, Dordrecht).

1977

'The Ethics of Convention and the Ethics of Nature', *Philosophic Exchange* **2**.

1978

Introductory Readings in Metaphysics, ed. by Richard Taylor with commentary and notes (Prentice-Hall, Inc., Englewood Cliffs, N.J.).

'Persons and Bodies', *American Philosophical Quarterly* **15**.

NOTES ON CONTRIBUTORS

D. M. ARMSTRONG, B. Phil., Ph.D., is Challis Professor of Philosophy in the University of Sydney. He is the author of *Berkeley's Theory of Vision, Perception and the Physical World, Bodily Sensations, A Materialist Theory of the Mind, Belief, Truth and Knowledge, Universals*, and numerous philosophical articles.

MYLES BRAND, Ph.D., wrote his doctoral thesis under the supervision of Richard Taylor. He is currently Professor of Philosophy at the University of Illinois at Chicago Circle. He is the editor of the books *The Nature of Human Action* and *The Nature of Causation* and co-editor of *Action Theory*. He has published numerous articles on various topics in metaphysics.

HÉCTOR-NERI CASTAÑEDA, Ph.D., is the Mahlon Powell Professor of Philosophy at Indiana University. He is at present (1980–81) the President of the Western Division of the American Philosophical Association and has held high office in many professional societies. His many books include *The Structure of Morality* and *Thinking and Doing: The Philosophical Foundations of Institutions*. He is the author of over a hundred articles and reviews, and is the founding editor of *Noûs*.

RODERICK M. CHISHOLM, Ph.D., is Andrew Mellon Professor of Humanities at Brown University. He has held visiting professorships and lectureships at numerous American and foreign universities, and is a fellow of the American Academy of Arts and Sciences. He is a past president of the Eastern Division of the American Philosophical Association and of the Metaphysical Society of America, and has held high office in numerous philosophical and educational organizations. In 1967 he gave the Carus Lectures. His books include *Perceiving: A Philosophical Study, Realism and the Background of Phenomenology* (ed.), *Theory of Knowledge*, and *Person and Object: A Metaphysical Study*. He is the author of over a hundred articles and reviews.

TIMOTHY DUGGAN, M.A., Ph.D., is Professor of Philosophy at Dartmouth College. He is the editor of Thomas Reid's *Inquiry Into the Human Mind*, and has published articles in *The Philosophical Review. The American Philosophical Quarterly, Noûs, The Encyclopedia of Philosophy*, and *Analysis and Metaphysics: Essays in Honor of Roderick M. Chisholm.*

JOEL FEINBERG, Ph.D., is Professor of Philosophy at the University of Arizona. He is the author of *Doing and Deserving: Essays in the Theory of Responsibility* and of numerous articles on moral philosophy, political philosophy, and the philosophy of law.

CARL GINET, Ph.D., is Professor of Philosophy at Cornell University. He is the author of *Knowledge, Perception, and Memory* and of articles in *The Philosophical Review, Noûs*, the *Canadian Journal of Philosophy, Acta Philosophica Fennica*, and in K.

Lehrer (ed.), *Freedom and Determinism*. (His articles include the classic 'Can the Will Be Caused?') He is a former editor of *The Philosophical Review*.

KEITH LEHRER, M.A., Ph.D., is Professor of Philosophy at the University of Arizona. He is the author of *Knowledge* and of many philosophical articles, and co-author of *Philosophical Problems and Arguments*. He is a co-editor of *Philosophical Studies*.

RAYMOND MARTIN, M.A., Ph.D., wrote his doctoral thesis under the supervision of Richard Taylor. He is currently Associate Professor of Philosophy at the University of Maryland, College Park. He is the author of articles in *Ratio*, *Philosophical Studies*, the *American Philosophical Quarterly*, *Theory and Decision*, *Philosophical Forum*, *Analysis*, and *Theoria*.

RICHARD M. MARTIN, M.A., Ph.D., is Professor of Philosophy at Northwestern University. He has held visiting professorships at several American and German universities, and is a member of the editorial boards of numerous journals. He is the author of ten books, including *Truth and Denotation*, *The Notion of Analytic Truth*, *Intension and Decision*, *Logic, Language, and Metaphysics*, and *Belief, Existence, and Meaning*.

MARGERY NAYLOR, M.A., Ph.D., has taught philosophy at Wells College and Syracuse University. She is the author of 'Chisholm on the Directly Evident'.

SYDNEY SHOEMAKER, Ph.D., is Professor of Philosophy at Cornell University. He is the author of the classic study of the problem of personal identity, *Self-Knowledge and Self-Identity* and of numerous articles on various topics in metaphysics. He is a former editor of *The Philosophical Review*.

J. J. C. SMART, M.A., B. Phil., is Professor of Philosophy in the Research School of Social Sciences of the Australian National University and Fellow of the Australian Academy of the Humanities. He has been a visiting professor at Harvard, Princeton, and Yale. He is the editor of the collection *Problems of Space and Time*, and the author of *Philosophy and Scientific Realism* and *Between Science and Philosophy*, and co-author of *Utilitarianism, For and Against*.

MARSHALL SWAIN, Ph.D., is Professor of Philosophy at the Ohio State University. He is the editor of the book *Induction, Acceptance, and Rational Belief* and co-editor of *Essays on Knowledge and Justification*. His papers have appeared in *The Journal of Philosophy*, *Synthese*, *Philosophical Studies*, *The Australasian Journal of Philosophy*, and the *American Philosophical Quarterly*.

IRVING THALBERG, Ph.D., is Professor of Philosophy at the University of Illinois at Chicago Circle. He is the author of *Enigmas of Agency* and *Perception, Emotion and Action: a Component Approach*. He is a contributor to the *Encyclopedia of Philosophy* and to many collections, including *Philosophy of Mind*, *Weakness of Will*, *Agent, Action, and Reason*, *Freud*, and *Mental Health: Philosophical Perspectives*.

PETER VAN INWAGEN, Ph.D., wrote his doctoral thesis under the supervision of Richard Taylor. He is currently Associate Professor of Philosophy at Syracuse University. He is the author of articles and reviews in the *American Philosophical Quarterly*, *Analysis*, *Noûs*, *Philosophia*, *The Philosophical Review*, *Philosophical Studies*, *Theoria*, and other journals. He is the author of a forthcoming book called *Free Will*.

INDEX

PHILOSOPHICAL STUDIES SERIES IN PHILOSOPHY

Editors:

WILFRID SELLARS, Univ. of Pittsburgh and KEITH LEHRER, Univ. of Arizona

Board of Consulting Editors:

Jonathan Bennett, Alan Gibbard, Robert Stalnaker, and Robert G. Turnbull

1. JAY F. ROSENBERG, *Linguistic Representation*, 1974.
2. WILFRID SELLARS, *Essays in Philosophy and Its History*, 1974.
3. DICKINSON S. MILLER, *Philosophical Analysis and Human Welfare*. Selected Essays and Chapters from Six Decades. Edited with an Introduction by Lloyd D. Easton, 1975.
4. KEITH LEHRER (ed.), *Analysis and Metaphysics*. Essays in Honor of R. M. Chisholm. 1975.
5. CARL GINET, *Knowledge, Perception, and Memory*, 1975.
6. PETER H. HARE and EDWARD H. MADDEN, *Causing, Perceiving and Believing*. An Examination of the Philosophy of C. J. Ducasse, 1975.
7. HECTOR-NERI CASTAÑEDA, *Thinking and Doing*. The Philosophical Foundations of Institutions, 1975.
8. JOHN L. POLLOCK, *Subjunctive Reasoning*, 1976.
9. BRUCE AUNE, *Reason and Action*, 1977.
10. GEORGE SCHLESINGER, *Religion and Scientific Method*, 1977.
11. YIRMIAHU YOVEL (ed.), *Philosophy of History and Action*. Papers presented at the first Jerusalem Philosophical Encounter, December 1974, 1978.
12. JOSEPH C. PITT, *The Philosophy of Wilfrid Sellars: Queries and Extensions*, 1978.
13. ALVIN I. GOLDMAN and JAEGWON KIM, *Values and Morals*. Essays in Honor of William Frankena, Charles Stevenson, and Richard Brandt, 1978.
14. MICHAEL J. LOUX, *Substance and Attribute*. A Study in Ontology, 1978.
15. ERNEST SOSA (ed.), *The Philosophy of Nicholas Rescher: Discussion and Replies*, 1979.
16. JEFFRIE G. MURPHY, *Retribution, Justice, and Therapy*. Essays in the Philosophy of Law, 1979.
17. GEORGE S. PAPPAS, *Justification and Knowledge: New Studies in Epistemology*, 1979.
18. JAMES W. CORNMAN, *Skepticism, Justification, and Explanation*, 1980.
19. PETER VAN INWAGEN, *Time and Cause*. Essays presented to Richard Taylor, 1980.
20. DONALD NUTE, *Topics in Conditional Logic*, 1980